Dynamics of Biological Macromolecules by Neutron Scattering

EDITED BY

Salvatore Magazù and Federica Migliardo

Department of Physics, University of Messina (Italy)

eBooks End User License Agreement

CONTENTS

FOREWORD

This e-book is devoted to applications of neutron scattering and other experimental methods to different biological problems, in which the dynamics of macromolecules plays the key role. Valuable information about the dynamics of proteins, membranes, lipids, nucleic acids and saccharides and so on can be found in this book. This information is very useful for many medical and pharmacological applications.

The book is arranged in the following way:

1) two papers from Section I by Mark T. F. Telling (1) and Salvatore Magazu *et al.* (2) are devoted to the brief discussion of those abilities which are given by the quasi-elastic neutron spectroscopy.

2) six other papers are divided into three Sections, which are devoted to the dynamics of macromolecules (Section II, papers (3) and (4)), the role of bioprotection mechanisms in extreme environment (Section III, papers (5), (6)) and the complementary aspects of the neutron spectroscopy, computer simulations and other techniques (Section IV, papers (7) and (8)).

It is necessary to note, that special attention in (1) is paid to the role of the intra-cellular water. It is manifested not only in the hydration effects. Its properties change with temperature and the cellular composition, and it its turn they exert influence on the dynamics of biomolecules. In order to clarify this circumstance, the study of the self-diffusion of water molecules is proposed in (1).

In (2) the main attention is focused on the investigation of the mean square displacement (MSD) of hydrogen containing fragments. In connection with this 1) the role of instrumental effects is analyzed and 2) peculiarities of the special procedure (using the Self Distribution Function) for the determination of MSD in different liquids and solutions are considered. Peculiarities of MSD for aqueous mixtures of two homologous disaccharides (i.e. sucrose and trehalose) and dry myoglobin in trehalose environment are discussed.

An overview of recent achievements in the investigation of biological membranes with the help of Elastic and Quasi-Elastic Incoherent Neutron Scattering is presented in (3). Here the main attention is paid to the study of the temperature dependence of MSD for bio-molecules of specified type in complex multicomponent systems, the wave vector dependence of the normalized intensity for EINS and the temperature dependence for the integrated elastic intensity of 1,2-Dimyristoyl-*sn*-Glycero-3-Phosphocholine. The last characteristics decrease essentially in the narrow temperature interval at $T \approx 315\,K$, that coincides with the temperature of the dynamic phase transition in pure water (*A. I. Fisenko and N. P. Malomuzh, Int. J. Mol. Sc., 10 (2009) 2383; Chem.Phys., 345 (2008) 164*). As seems to us such a coincidence is scarcely occasional.

Some questions of the dynamics of Myoglobin in a confined geometry are discussed in the paper (4). This geometry is realized by encapsulation of Myoglobin in a porous silica matrix. Using elastic neutron scattering allows to study the temperature dependence of the MSD of non-exchangeable hydrogen atoms. It is demonstrated that geometrical confinement plays a crucial role in protein dynamics. Mechanisms of suppression of cooperative relaxation in confinement systems are investigated with the help of dielectric spectroscopy.

The bio-protection function of saccharides is analyzed in the paper (5). For this aim vibrational properties of proteins embedded in amorphous saccharide matrices are considered. The experimental results obtained with the help of FTIR and SAXS are completed by computer simulation data. Due to this, different efficiency of saccharides and carbohydrates in preserving biostructures becomes quite clear. The comparison among various sugars matrices shows why saccharides behave differently in their interaction with water and bio-molecules, why trehalose has better biopreserving properties.

The modification of the morphology of lipid assemblies and their dynamics by additions of saccharides are discussed in the review (6). It is taken into account that sugars influence essentially the behavior of headgroups of lipid molecules and methyl groups of their hydrocarbon chains. As a result, significant decrease of the temperature for gel to liquid–crystalline phase transition at the addition of saccharides is observed. The importance of this question is connected with the necessity to maintain cell viability in stressful environmental conditions. The influence of saccharides is studied with the help of DSC, SAXS FTIR and NMR technique. A fine tuning of the multilayer structure at varying the amount of sugar and the stabilization of lipid multilayers is demonstrated. Influence of charged polyelectrolytes on clusterization and the self-assembling in colloidal systems is discussed.

The possibilities of the neutron vibrational spectroscopy or inelastic neutron scattering (INS) are considered in the paper (7). It gives us a clear representation about INS as powerful instrument for receiving significant information about vibrational modes of bio-molecules at non-zero wave vectors, which is complementary to that of Raman and Infrared Spectroscopy. In contrast with the last, INS is sensitive to hydrogen motion. Several characteristic examples for the application of INS are presented. In particular, the spectral function $S(\vec{Q},\omega)$ and the dispersion curve for the vibrations of DNA are given. Special attention is paid to the manifestations of infra-cellular water, in particular, influence of the hydration effects and isotopic replacement of H on D.

A synergistic aspect of MD simulations and experimental information obtained by inelastic neutron scattering, small angle X-ray and neutron scattering, and other techniques is discussed in the paper (8). Comparison of different approaches is realized at the level of the scattering functions or quantities derived from them. The situation is clearly illustrated by a schematic square. The usefulness of such an approach to the analysis of protein dynamics is demonstrated at the consideration of fine details in the behavior of protein chains near the glass transition temperature. This methodology is especially important for the testing of simulation models, applied, in particular, to the study of the protein-solvent dynamical coupling and hydration effects.

This book not only introduces new results obtained by neutron scattering methods, but it also stimulates the formulation of new ideas. I hope that this book will be useful for readers of different fields: specialists in Biophysics and its applications, medics, physical, biological and medical students and postgraduate students.

Professor Nikolay P. Malomuzh

PREFACE

The scope of the E-book entitled "Dynamics of Biological Macromolecules by Neutron Scattering" is to provide insight into the study of the dynamics of biological macromolecules by neutron scattering techniques.

The E-book is focused on recent scientific results on biomolecular motions obtained by using neutron spectroscopy also in combination with simulative methods and complementary spectroscopic techniques and reflects the importance of progresses and innovation which characterise today this kind of studies.

Neutron scattering is as a powerful tool to find answers to a wide range of interesting scientific questions. There is a broad consensus that in the near future the potential for innovative work in neutron scattering is greatest in the area of biophysics for two main reasons: (i) experimentally, bringing sample sizes down by an order of magnitude will open up a rich field of work on interactions in functionally important systems; (ii) interpretationally, multi-parameter data sets of high quality allowing more detailed comparisons with increasingly realistic Molecular Dynamics simulations of biomolecules and their building blocks.

At present, an increasing number of physicists study, by new transdisciplinary approaches, systems of biological interest where complexity reigns, and are discovering how rewarding the interaction with biologists can be. In the living world, complexity implies a degree of organizational hierarchy, defined by several length scales with an interplay between events at different levels. This interplay extends from the events that happen very slowly on a global scale right down to the most rapid events observed on a microscopic scale.

The E-book discusses likely directions of future work on biological macromolecular systems and outlines some challenging, hitherto inaccessible problem areas.

The applicability of neutron scattering to everwidening fields of biological studies is more and more extended and neutron scattering community is interested on using the unique capabilities of different facilities and technologies to their best advantage. Furthermore the large neutron fluxes produced in next-generation spallation facilities are likely to soon find expanded application in biology.

CONTRIBUTORS

Antonio Benedetto

Dipartimento di Fisica, Università di Messina, P.O. Box 55, I-98166 Messina, Italy

Lorenzo Cordone

Dipartimento di Scienze Fisiche ed Astronomiche, Università di Palermo, I-90123, Palermo, Italy

Grazia Cottone

Dipartimento di Scienze Fisiche ed Astronomiche, Università di Palermo, I-90123, Palermo, Italy and School of Physics, University College Dublin, Dublin, Ireland

Antonio Cupane

University of Palermo, Dept. of Physical and Astronomical Sciences, Italy

Antonio Deriu

Dipartimento di Fisica, Università degli Studi di Parma and CNISM , Viale G.P. Usberti 7/a, Parma, 43100, Italy

Maria Teresa Di Bari

Dipartimento di Fisica, Università degli Studi di Parma and CNISM , Viale G.P. Usberti 7/a, Parma, 43100, Italy

Yuri Gerelli

Dipartimento di Fisica, Università degli Studi di Parma and CNISM , Viale G.P. Usberti 7/a, Parma, 43100, Italy

Sergio Giuffrida

Dipartimento di Scienze Fisiche ed Astronomiche, Università di Palermo, I-90123, Palermo, Italy

Miguel A. Gonzalez

Institut Laue Langevin, 6, Rue Jules Horowitz, F-38042 Grenoble Cedex 9, France

Marimuthu Krishnan

UT/ORNL Center for Molecular Biophysics, Oak Ridge National Laboratory, Oak Ridge, TN 37831, USA

Alessandro Longo

CNR-ISMN, I-90146, Palermo, Italy

Salvatore Magazù

Dipartimento di Fisica, Università di Messina, P.O. Box 55, I-98166 Messina, Italy

Federica Migliardo

Dipartimento di Fisica, Università di Messina, P.O. Box 55, I-98166 Messina, Italy

Claudia Mondelli

CNR-INFM-OGG and CRS Soft, Institut Laue Langevin, 6, Rue Jules Horowitz, F-38042 Grenoble Cedex 9, France

Francesca Natali

CNR-INFM, OGG, c/o Institut Laue-Langevin, Grenoble, FR-38000, France

Stewart F. Parker

ISIS Facility, STFC Rutherford Appleton Laboratory, Chilton, Didcot, Oxfordshire, OX11 0QX, UK

Loukas Petridis

UT/ORNL Center for Molecular Biophysics, Oak Ridge National Laboratory, Oak Ridge, TN 37831, USA

Giorgio Schirò

University of Palermo, Dept. of Physical and Astronomical Sciences, Italy

Jeremy C. Smith

UT/ORNL Center for Molecular Biophysics, Oak Ridge National Laboratory, Oak Ridge, TN 37831, USA

Nikolai Smolin

UT/ORNL Center for Molecular Biophysics, Oak Ridge National Laboratory, Oak Ridge, TN 37831, USA

Mark T. F. Telling

ISIS Facility, Rutherford Appleton Laboratory, OX11 OQX, UK

Marcus Trapp

Institut de Biologie Structurale J.-P. Ebel, UMR 5075, CNRS-CEA-UJF, Grenoble, FR-38027, France

SECTION I

Neutron Scattering as a Powerful Tool for Studying Biological Molecules and Processes

Quasi-Elastic Neutron Scattering – A Tool for the Study of Biological Molecules and Processes

Mark T.F. Telling[*]

ISIS Facility, Rutherford Appleton Laboratory, OX11 OQX, UK

Abstract: The spatial and temporal ranges accessible using the technique of quasi-elastic neutron scattering (QENS) are ideally matched to the atomic and molecular vibrational displacements, correlation lengths and diffusive motions encountered in highly complex biological systems. The QENS method has been successfully applied to a diverse range of bio-molecular problems which encompass, for example, proteins, membranes, lipids, nucleic acids and saccharides. In this section, the basic principles of quasi-elastic neutron scattering pertinent to the study of dynamic processes in biological molecules are presented. An overview of the neutron instrumentation required for such studies is given as are experimental results which highlight the ideas outlined.

INTRODUCTION

Neutron scattering has helped underpin our current understanding of soft matter science by enabling scientists to answer two fundamental questions - 'where atoms are' and 'what atoms do'. Answers to such elemental questions are possible because of the unique properties of the neutron. Neutron sources [1] produce beams of neutrons whose wavelengths not only span the inter-atomic spacing of complex biological molecules, but also whose energies are comparable to vibrational frequencies and the diffusive motions of the molecular components. More importantly, the neutron is a highly penetrating, weakly perturbing, yet non-destructive, probe. This is extremely important for the study of biological macromolecules where samples can have limited quantity and a high cost. Indeed, the complementarities of neutron scattering with other experimental methods (see Fig. 1) have also greatly advanced our understanding of bio-related topics; the neutron allowing access to length and time scales inaccessible using, for example, light scattering techniques alone. However, it is the ability to selectively 'label', using deuterium, different molecular components that is key for the study of highly complex, hydrogenous biological systems using neutrons. Deuterium labelling allows the neutron spectroscopist to discriminate between, and better understand, different dynamic processes. So important is the deuteration technique that neutron facilities embrace deuteration support facilities; examples include the ILL-EMBL Deuteration Facility (D-lab) at the Institut Laue Langevin (ILL), France [1] and the National Deuteration Facility (NDF) [2] co-funded by the Australian Nuclear Science and Technology Organisation (ANSTO).

Quasi-elastic neutron scattering (QENS) is sensitive to the re-organisation of atoms and molecules on a pico-second (ps) to nano-second (ns) time scale (10^{-13} to 10^{-7} seconds), with energies 10^4 to 10^{-2} micro-electronvolt (μeV) over length scales 1 to ~ 500 Angstrom (Å) ; length scales which cover both inter and intra molecular distances. Typically, a quasi-elastic neutron scattering event is measured using time-of-flight (t.o.f) direct geometry, t.o.f. indirect geometry (also known as 'backscattering'), non-t.o.f indirect geometry and/or neutron spin echo neutronics. As Fig. **1** demonstrates, a combination of these neutron techniques is required to access the full dynamic window.

This broad spatial and temporal scale is ideally matched to the atomic and molecular vibrational displacements, jump distances and correlation lengths encountered in highly complex biological systems. These systems display extraordinary structural and functional diversity and it is the inter-relation of structure, dynamics and function at the molecular level that is a central theme of molecular biology. Short range translational and rotational diffusion,

---

[*]**Address correspondence to Mark T.F. Telling:** ISIS Facility, Rutherford Appleton Laboratory, OX11 OQX, UK; E-mail: mark.telling@stfc.ac.uk
[1]Institut Laue-Langevin (ILL), France (www.ill.eu) ; Australian Nuclear Science and Technology Organisation (ANSTO), Australia (www.ansto.gov.au); Helmholtz Zentrum Berlin (HZB), Germany (www.helmholtz-berlin.de); Japan Proton Accelerator Research Complex (J-PARC), Japan (www.j-parc.jp); Forschungs-Neutronenquelle Heinz Maier-Leibnitz (FRM II), Germany (www.frm2.tum.de); Paul Scherrer Institut (PSI), Switzerland (sinq.web.psi.ch); NIST Centre for Neutron Research (NCNR), USA (www.ncnr.nist.gov); Laboratoire Léon Brillouin (LLB), France (www-llb.cea.fr); Spallation Neutron Source (SNS), USA (http://neutrons.ornl.gov/); ISIS, UK (http://www.isis.stfc.ac.uk/)"

Salvatore Magazù and Federica Migliardo (Eds)

alongside low-frequency vibrational modes, are believed essential for biological function [3]. Furthermore, the inherent presence of water is a fundamental prerequisite for biological efficacy.

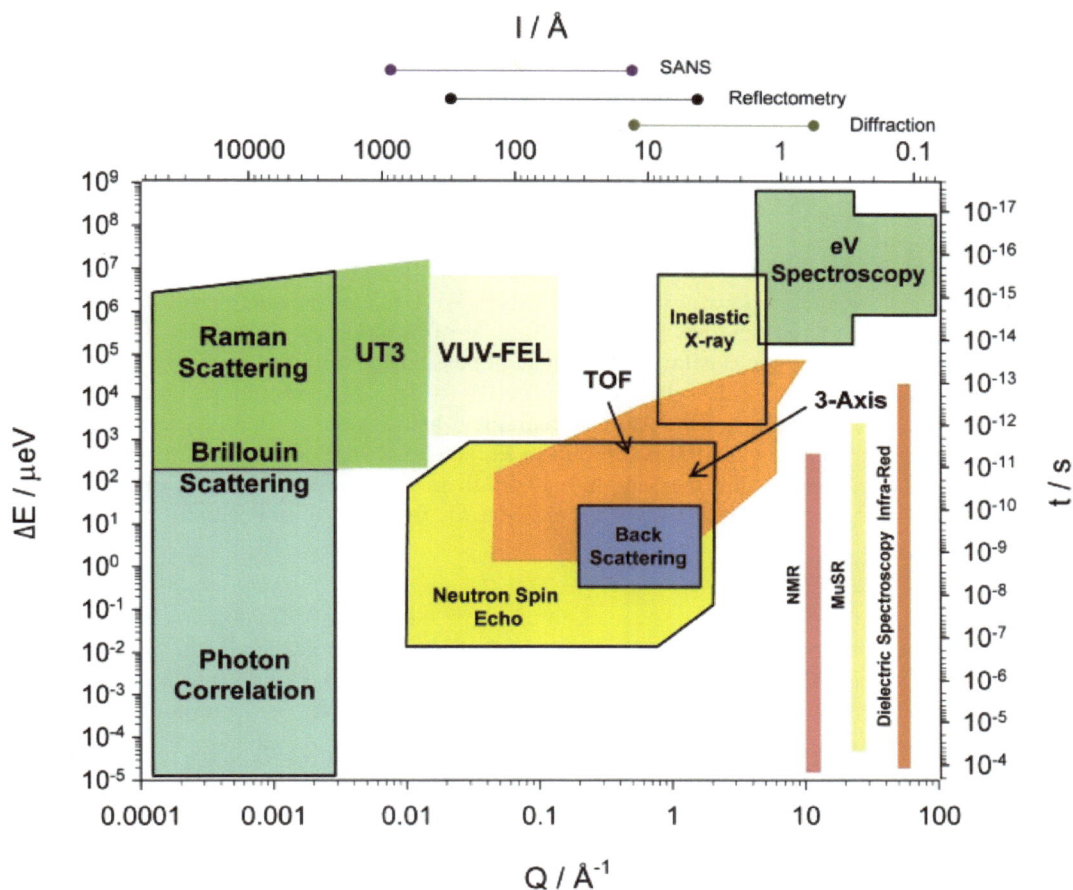

Figure 1: Complementarity of neutron scattering tecnhiques and other experimental methods. The diagram correlates length (l), time (t), energy (ΔE) and momentum transfer (Q) scales. Techniques that do not directly provide distance information are indicated as bars along the time axis. Future neutron (The European Spallation Source [4]) and x-ray instrumentation (The European X-Ray Laser Project, XFEL [5]) should bridge the gaps that exist at present bewteen the techiques.

While natural bio-material exhibits a certain degree of conformational freedom, an aqueous environment serves to further enhances the landscape [6, 7]; a hydration layer, for example, can provide room for those additional conformational degrees of freedom necessary for bioactivity. A clear understanding of biomolecule-water interactions is therefore of paramount importance. It is also worth mentioning that, while beyond the scope of this section, the temporal and spatial ranges accessible using Molecular Dynamic (MD) simulations overlap those probed by QENS. The QENS method is thus an excellent experimental tool with which to validate, and advance, MD dynamic simulations of complex biological macromolecules [8, 9].

The range of bio-macromolecular problems addressed using quasi-elastic neutron spectroscopy is considerable. For a comprehensive overview see for example Fitter *et al.* [10], García Sakai *et al.* [11] and references therein. However, in broad terms the QENS method has been successfully applied to problems that encompass proteins [12], membranes [13, 14], lipids [15, 16], nucleic acids [17] and saccharides [18-20]. It is also worth mentioning that the QENS method is also aiding the development of bio-compatible polymer-based drug delivery materials [21,22]. These materials have the potential to deliver biologically active cargo (e.g. a DNA fragment) to specific organs.

Neutron scattering is a well established technique. As a result, comprehensive details of the neutron scattering method and its applications can be found in [23-25]. In this chapter, the basic principles of quasi-elastic neutron scattering pertinent to the study of dynamic processes in biological macromolecules are presented. An overview of neutron instrumentation is also given. The chapter concludes with experimental results that highlight the ideas outlined.

QUASI-ELASTIC NEUTRON SCATTERING (QENS)

The technique of quasi-elastic neutron scattering concerns itself with the measure of low-energy spectra associated with diffusive or vibrational atomic and molecular motion [23-25]. The QENS technique is sensitive to those neutron scattering events which involve small changes of energy (typically $\pm$ 2000 μeV) between the incident neutron and scattering centre (*i.e.* an atomic nucleus). This change of energy, referred to as an energy transfer (ΔE), results in a peak in detected neutron intensity of finite width, Γ, centred about the energy transfer equals zero position. The intensity and breath of this peak can exhibit marked angular dependence. It is by modelling these parameters as a function of, for example, temperature, pressure and / or composition that macromolecular dynamics can be better understood. The energy transfer itself, ΔE ($\hbar\omega = E_i - E_f$), relates the energy of the incident neutron, E_i, to the energy of the same scattered neutron, E_f. The magnitude of the momentum transfer, Q ($= (4\pi\lambda^{-1})\sin(\theta/2)$ where λ is the neutron wavelength and θ is the scattering angle) associated with the scattering event is proportional to the neutron wave vectors before, k_i ($= 2\pi/\lambda_i$) and after, k_f ($= 2\pi/\lambda_f$). It should be noted that if the energy exchange is smaller than the resolution of the neutron instrument being used, or equal to zero, then the scattering process is termed *elastic* (the position $\Delta E = 0$ is referred to as the elastic line). In contrast, a non-zero exchange of energy that leads to an excitation within the sample may result in an inelastic scattering event. Experimentally, inelastic scattering manifests itself as a peak, or peaks, centred away from the $\Delta E = 0$ position.

The Scattering Function

A neutron scattering experiment is essentially a measurement of the double differential scattering cross-section, $d^2\sigma/(d\omega d\Omega)$ [23] *i.e.* the probability that a neutron of incident energy E_i is scattered into a solid angle element $d\Omega$ about the direction Ω and with a change of energy between $\Delta E = \hbar\omega = E_f - E_i$ and $\hbar(\omega + d\omega)$. The detected neutron intensity is analysed as a function of energy and momentum transfer, and proportional to the so-called scattering function $S(Q,\omega)$. As a result,

$$\frac{d^2\sigma}{d\Omega d\omega} \propto \frac{k_f}{k_i}\frac{\sigma}{4\pi}S(\mathbf{Q},\omega)$$ (1)

where σ is the total scattering cross section of the bound scattering centre. Also known as the dynamic structure factor, analysis of $S(Q,\omega)$ can reveal aspect of the dynamic landscape within a biological material.

The Importance of the Hydrogen Atom

Before examining the information contained within $S(Q,\omega)$ it is worth considering the relevance of the scattering cross section, σ, and the influence of the hydrogen and deuterium atom.

When a neutron passes near a nucleus it will either be scattered or absorbed. The scattering cross section, σ, therefore enables the relative number of scattered and absorbed bodies to be determined. Considering a mono-atomic sample, if I_i neutrons hit the sample per cm^2 per second then the number of neutrons scattered / adsorbed can be written $I_s = \sigma_s I_i$ and $I_a = \sigma_a I_i$ respectively. Here σ_s and σ_a are the scattering and absorption cross sections. Both quantities have the dimension *barns* (x 10^{-24} cm^2) and are tabulated in full, in [26]. Note: for incident neutron energies less than a few milli-electronvolts (meV) $\sigma_a \propto \lambda_n$. In [26] σ_a is quoted for 1.8 Å neutrons.

Further comment should be made about the scattering cross section itself. The neutron-nucleus interaction is isotropic and characterized by an energy independent parameter, b; the scattering length. The scattering length can be complex. While the imaginary component represents absorption, the real component will assume a + ve or – ve value depending upon the attractive or repulsive nature of the neutron-nucleus interaction. The scattering process therefore depends on both the nature of the nucleus but also the total spin state of the neutron-nucleus interaction. A mono-atomic system comprised of a mixture of different isotopes each with a nuclear spin will result in a random distribution of scattering lengths. Compared to the response observed from a system of identical nuclei with the same scattering length, a random distribution of b serves to alter interference effects between scattered neutron waves and thus modify the detected neutron intensity. More complex systems, such as biological macromolecules, also comprise of n different types of atom. Each atom will exhibit its own isotope and spin mix and thus further modify the interference phenomena. As a result, the scattering cross section must include both coherent and incoherent interference contributions ($\sigma = \sigma_{coh} + \sigma_{inc}$). While the former is associated with terms arising from constructive interference between pairs of atoms, the latter denotes interference phenomena which cancel out. As a result the scattering function becomes,

$$\frac{d^2\sigma}{d\Omega d\omega} = \frac{k_f}{k_i 4\pi N}\left[\sigma_{coh}S_{coh}(\mathbf{Q},\omega) + \sigma_{inc}S_{inc}(\mathbf{Q},\omega)\right] \qquad (2)$$

For a system comprised of i different scattering lengths, $b_1...b_i$, with concentrations, c_1 to c_i, the coherent and incoherent scattering cross sections can be related to b via: $\sigma_{coh} = 4\pi \overline{b}^2$ (where $\overline{b} = \Sigma c_i b_i$) and $\sigma_{inc} = 4\pi(\overline{b^2} - \overline{b}^2)$ (where $\overline{b^2} = \Sigma c_i b_i^2$) [10]. The intensity of the scattered beam contains both incoherent and coherent scattering contributions weighted according to the magnitude of the corresponding scattering cross sections. Weighting has important consequences for the study of complex multi-component bio-materials. An advantage of using neutrons for the study of biological molecules is that one of the most common nuclei is hydrogen (H). Hydrogen has a sizable incoherent scattering cross section compared to other nuclei, σ^H_{inc} = 79.9 barns. Indeed, σ^H_{inc} is 40 times greater than its own coherent cross section, σ^H_{coh} = 1.8 barns. $S(Q,\omega)$ measured from hydrogenous (also referred to as protonated) material is therefore dominated by the incoherent scattering signal arising from the H atoms. Of greater benefit, however, is the fact that the hydrogen isotope, deuterium, has a much weaker incoherent cross section, σ^D_{inc} = 2.04 barns. By replacing with deuterium those hydrogen atoms associated with certain molecular species (*i.e.* perhaps a particular amino acid) the incoherent signal, and hence dynamic response, associated with particular chemical groups can be masked. For complex systems such as biological macromolecules, this technique (also referred to as 'labeling' or 'selective deuteration') allows different dynamic processes to be isolated.

Interpreting S (Q,ω)

The form of the coherent and incoherent scattering terms was derived by Van Hove [27] *via* the energy independent static structure factor, S(Q). Van Hove showed that,

$$S_{coh}(\mathbf{Q},\omega) = \frac{1}{2\pi}\int_{-\infty}^{+\infty} I(\mathbf{Q},t)e^{-i\omega t}dt \text{ and } S_{inc}(\mathbf{Q},\omega) = \frac{1}{2\pi}\int_{-\infty}^{+\infty} I_{inc}(\mathbf{Q},t)e^{-i\omega t}dt \qquad (3)$$

where I(Q,t) and I_{inc}(Q,t) represent intermediate scattering functions. In addition, Van Hove was able to correlate the positions of nuclei, either collectively or individually, after a given time, t, *via* the so-called pair and self correlation functions, G(r,t) and G_{inc}(r,t). As a result,

$$S_{coh}(\mathbf{Q},\omega) = \frac{1}{2\pi}\int_{-\infty}^{+\infty}\int_{-\infty}^{+\infty} G(\mathbf{r},t)\exp^{-i(\mathbf{Qr}-\omega t)} d\mathbf{r}dt \qquad (4)$$

and

$$S_{inc}(\mathbf{Q},\omega) = \frac{1}{2\pi}\int_{-\infty}^{+\infty}\int_{-\infty}^{+\infty} G_{inc}(\mathbf{r},t)\exp^{-i(\mathbf{Qr}-\omega t)} d\mathbf{r}dt \qquad (5)$$

Using classical approximations and meanings for the intermediate scattering and correlation functions, $S_{inc}(Q,\omega)$ and $S_{coh}(Q,\omega)$ can be interpreted as follows. Incoherent scattering is a measure of self-correlation of single atoms *i.e.* correlations between the positions of the same atom at different times. Dynamics associated with incoherent scattering are thus from independent nuclei and related to self motion. In contrast, coherent scattering is concerned with atom-atom pair-correlations *i.e.* correlations between the positions of different atoms at different times. The resulting $S_{coh}(Q,\omega)$ is therefore a measure of collective processes. Interpretation of the correlation functions in the classical limit is valid for small energy and momentum changes; quantum effects may be expected over small distances and large ω. As a result, $S(Q,\omega)$ derived above using a classical interpretation should be corrected by the so-called detailed balance factor, $\exp(-\hbar\omega / 2k_BT)$.

Due to the high hydrogen content in biological materials QENS is, in general, a measure of the incoherent scattering function and thus self motion. From here on we shall only concern ourselves with interpretation of $S_{inc}(Q,\omega)$.

Spectral Contributions to $S_{inc}(Q,\omega)$

Consider a bulk sample comprised of identical molecules. If only one dynamically equivalent atom type dominates the scattering process then the incoherent scattering law resorts to the case of a single scatterer [23]. The spectroscopic contributions to $S_{inc}(Q,\omega)$ are therefore,

$$S_{inc}(\mathbf{Q},\omega) = S_{inc}^{L}(\mathbf{Q},\omega) \otimes S_{inc}^{R}(\mathbf{Q},\omega) \otimes S_{inc}^{V}(\mathbf{Q},\omega) \tag{6}$$

where $S_{inc}^{L}(Q,\omega)$ is a contribution from lattice modes, $S_{inc}^{V}(Q,\omega)$ represents intra-molecular vibrations and $S_{inc}^{R}(Q,\omega)$ signifies molecular reorientations. $\otimes$ is the convolution product. For a liquid, $S_{inc}^{L}(Q,\omega)$ should be replaced by $S_{inc}^{T}(Q,\omega)$ to include long-range translational diffusion. It can be show that both $S_{inc}^{L}(Q,\omega)$ and $S_{inc}^{V}(Q,\omega)$ comprise of both elastic and in-elastic components. For example, for the latter,

$$S_{inc}^{V}(\mathbf{Q},\omega) = \exp(-<u_V^2>\mathbf{Q}^2)[\delta(\omega) + S_{inel}^{V}(\mathbf{Q},\omega)] \tag{7}$$

Here $(<u_V^2>Q^2)$ is the Debye-Waller factor (DWF) and $<u_V^2>$ represents the mean square displacement (msd) of the atoms due to molecular vibrations. $S_{inc}^{L}(Q,\omega)$ assumes the same form as above but with $<u_V^2>$ replaced by $<u_L^2>$; the latter being a measure of the atomic msd associated with lattice modes. Both $S_{inc}^{L}(Q,\omega)$ and $S_{inc}^{V}(Q,\omega)$ have little affect of the quasi-elastic form of $S_{inc}(Q,\omega)$. In brief, $S_{inc}^{V}(Q,\omega)$ manifests itself as in-elastic spectral lines peaked at high energy transfer. $S_{inc}^{L}(Q,\omega)$, in contrast, takes the form of a small background in the quasi-elastic regime. As a result, the scattering law for a bulk sample can be simplified,

$$S_{inc}(\mathbf{Q},\omega) = \exp(-<u^2>\mathbf{Q}^2)[S_{inc}^{R}(\mathbf{Q},\omega) + S_{inc}^{I}(\mathbf{Q},\omega)] \tag{8}$$

Here $<u^2> = <u_V^2> + <u_L^2>$. $S_{inc}^{I}(Q,\omega)$ is an illustrative in-elastic contribution which, again, contributes little to the quasi-elastic regime. For a sample exhibiting long-range translational motion, $S_{inc}^{R}(Q,\omega)$, in (8) should be replaced by, $S_{inc}^{R}(Q,\omega) \otimes S_{inc}^{T}(Q,\omega)$ and $<u^2>$ by $<u_V^2>$.

Basic Characteristics of the Incoherent Scattering Function

Modeling, and subsequent interpretation, of the rotational, $S_{inc}^{R}(Q,\omega)$, and / or translational, $S_{inc}^{T}(Q,\omega)$, incoherent scattering functions extracted from a QENS measurement is a basic goal of any QENS experiment. While the exact spectral responses depend upon the sample under investigation and external environmental factors, there are certain characteristics which define the two scattering functions. These characteristics become apparent if one considers the effect of the self correlation function at infinite time. Consider two atoms. The first diffuses within a space that is very large compared to the inter-atomic distance. The second is constrained to a move within a small finite volume. As $t \to \infty$ the self correlation function for the former, $G_{self}(r, \infty)$ tends to zero. In contrast, for constrained movement, $G_{self}(r, t)$ tends towards a finite, time-independent, value. If the self correlation function is split into these time-dependent and time-independent terms, $G_{self}(r, t) = G_{self}(r, \infty) + G'_{self}(r, t)$, then it can be shown that the incoherent scattering function takes the form,

$$S_{inc}(\mathbf{Q},\omega) = I_{inc}^{el}(\mathbf{Q},\infty)\delta(\omega) + S_{inc}^{qe}(\mathbf{Q},\omega) \tag{9}$$

i.e. the sum of purely elastic and quasi-elastic components. Scattering from a dynamically disordered system is characterized by the absence of the elastic component since $G_{self}(r, \infty)$ vanishes. In contrast, existence of an elastic contribution signals the presence of a scattering center, such as a rotating side molecule on a peptide chain, constrained in space. For both cases, the width of $S^{qe}_{inc}(Q,\omega)$ contains information about the characteristic time of the associated motion. However, for the latter, the relative integrated intensities of the elastic, $I^{el}(Q)$, and quasi-elastic, $I^{qe}(Q)$, components also contain information about the geometry of the constrained motion. Such information is revealed *via* the so-called Elastic Incoherent Structure Factor (EISF), $A_o(Q)$,

$$ EISF = A_o(\mathbf{Q}) = \frac{I^{el}(\mathbf{Q})}{I^{el}(\mathbf{Q}) + I^{qe}(\mathbf{Q})} \tag{10} $$

A momentum transfer dependent EISF denotes a specific type of motion localized in space and careful extraction, and analysis, of the EISF using theoretical models can reveal the associated geometry. As an example, Fig. **2** shows the EISF response expected for an atom jumping between 2, 3 and 4 sites equally spaced on a circle of radius, r = 1 Å. Experimentally determined EISF's associated with other geometries, length scales and frequencies are compared to theoretical predictions in [23]. Ideally, the elastic and quasi-elastic spectral integrals are evaluated for $\hbar\omega = \pm \infty$. If so, the sum of the relative intensities will be Q independent. As a result, Eq. (9) can be written in terms of elastic and quasi-elastic incoherent structure factors,

$$ S_{inc}(\mathbf{Q},\omega) = A_o(\mathbf{Q})\delta(\omega) + A_1(\mathbf{Q})L(\omega) \tag{11} $$

Here, the quasi-elastic component is described using a Lorentzian function, $L(\omega)$. However, since no neutron instrument straddles such an infinite dynamic window, care must be taken when evaluating the EISF experimentally.

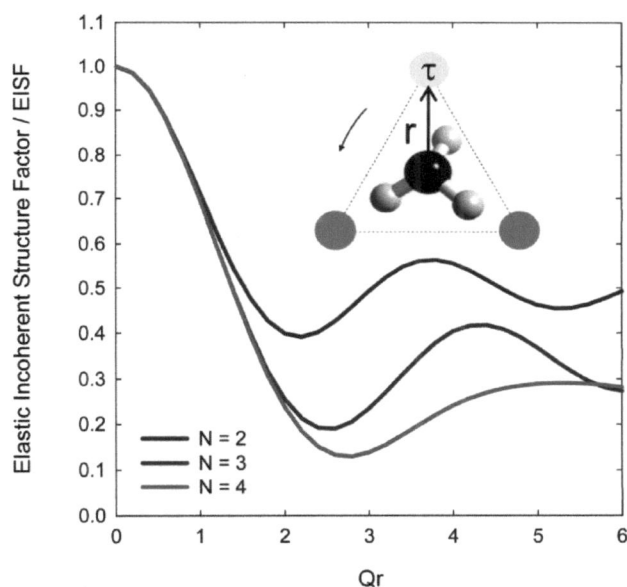

Figure 2: Theoretical EISF curves for atoms jumping between 2, 3 or 4 sites equally spaced on a circle of radius, r (= 1 Å). The cartoon represents a three-site, 120° jump rotation model; a model used successfully to describe the reorientation of methyl (CH_3) groups. τ is the residence time at each site.

QENS SPECTROMETERS

Several types of neutron spectrometer exist for the study of quasi-elastic neutron scattering in biological macromolecules [28]. Design-comparable spectrometers include direct-geometry [2] and indirect geometry [3] neutron

---

[2] **NEAT** (HZB) [29]; **TOF-TOF** (FRM-II) [30]; **IN5** (ILL) [31]; **IN6** (ILL) [32]; **FOCUS** (PSI) [33]; **DCS** (NCNR) [34]; **MIBEMOL** (LLB) [35]; **AMATERAS** (formerly CNDCS, J-PARC) [36]; **PELICAN** (ANSTO) [37]; **LET** (ISIS) [38]; **CNCS** (SNS) [39]

instruments. In addition, the neutron spin echo[4] (NSE) spectrometer can be used to investigate QENS processes. Examples of each type, and links to their respective instrument characteristics, are given in the footnotes. It should be mentioned that while the former evaluates the quasi-elastic response in terms of energy, $S_{inc}(Q,\omega)$, the latter (NSE) provides a direct measure of the time-dependent intermediate scattering function, $I(Q,t)$. As Fig. 3 illustrates, these spectrometer types allow access to a time regime from 10^{-13} to 10^{-7} seconds and probe length scales from ~ 1 to ~ 500 Å. Clearly, no neutron instrument alone straddles the complete temporal range. Indeed, no one instrument has suitably fine energy resolution as to discriminate between all dynamics processes. As a result complete understanding of a macromolecule's dynamic landscape requires data from different neutron instruments, and perhaps experimental techniques, to be collated. This approach of course has inherent experimental limitations. Nonetheless, the choice of spectrometer is highly dependent upon the motion to be observed and thus the temporal and spatial resolution required.

The Direct Geometry Spectrometer

To determine ΔE, accurate knowledge of either the incident, or scattered, neutron energy is required. A direct geometry spectrometer see Fig. 3 operates using a highly monochromatic incident neutron beam of energy, E_i and employs the time-of-flight (t.o.f) technique to determine the final energy of each scattered neutron, E_f. The t.o.f method is essentially a technique by which the arrival of a scattered neutron at the detector is time-stamped by the detection electronics; this time stamp being relative to a well defined t = 0. All spectrometer distances are accurately known and the detector is sufficiently far away from the sample (order of meters) to allow the t.o.f of the scattered neutron to be ascertained. The measured t.o.f can be related to the energy exchanged within the sample. A sizable amount of spatial, or Q, information can be collected in a single measurement by positioning a multi-detector assembly radially about the sample position. The direct geometry technique can be further sub-divided depending upon the method used to monochromate the incident neutron beam. Two techniques are commonly applied. The first employs an array of velocity-selectors (an arrangement of narrow-aperture disc or Fermi choppers) which allow neutrons of only a certain wavelength to reach the sample position. The second method utilizes the principle of Bragg scattering from an array of single crystals to define, E_i. Typically, the direct geometry spectrometer probes the pico-second time regime and has an energy transfer resolution in the meV regime.

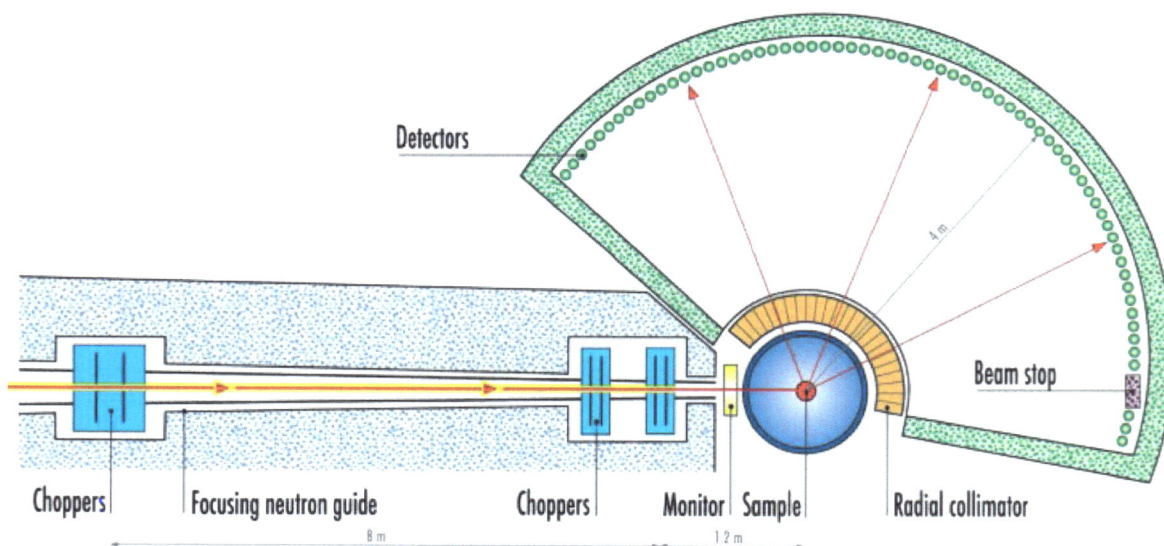

Figure 3: An example of a direct geometry time of flight instrument. IN5 at the ILL, France [31] uses a chopper system to monochromate the incident beam. The figure is taken from [59] and is courtesy of the ILL

---

[3] **HFBS** (NCNR) [40]; **OSIRIS** (ISIS) [41]; **IRIS** (ISIS) [42]; **DNA** (formerly DYANA, J-PARC); **BASIS** (SNS) [43]; **IN10** (ILL) [44], **IN16** (ILL) [45]; **IN16b** (ILL) [46, 47]; **IN13** (ILL) [48]; **MARS** (PSI) [49]; **SPHERES** (FRM-II) [50]
[4] **IN11A / C** (ILL) [50]; **IN15** (ILL) [51-53]; **NSE** (NIST) [54]; **JNSE** (JCNS) [55]; **RESEDA** (FRM-II) [56]; **iNSE** (formerly ISSP-NSE, ISSP) [57], **WASP** (ILL) [58]

The Indirect Geometry Spectrometer

In contrast, the indirect geometry instrument operates using a distribution, ΔE_i, of incident neutron wavelengths and a well defined final, or detected, neutron energy, E_f; Bragg scattering being used to energy analyze the scattered neutron beam. Energy analysis is achieved by placing a large array of single crystals (typically silicon ([111] reflection) or pyrolytic graphite ([002] reflection)) in the path of the scattered beam between sample and detector. Only those elastically scattered, or energy exchanged, neutrons that leave the sample with a wavelength that satisfies the Bragg condition will be directed towards the detector. Indirect geometry spectrometers are also known as *backscattering* instruments since, to achieve the highest energy resolution, the energy analyzed beam is Bragg reflected back ($2\theta \rightarrow 180^\circ$) towards detectors located behind, or slightly below, the sample position. In general, all backscattering instruments straddle similar Q ranges and use similar energy analysis techniques to isolate E_f. However, the process of generating a distribution in E_i can differ.

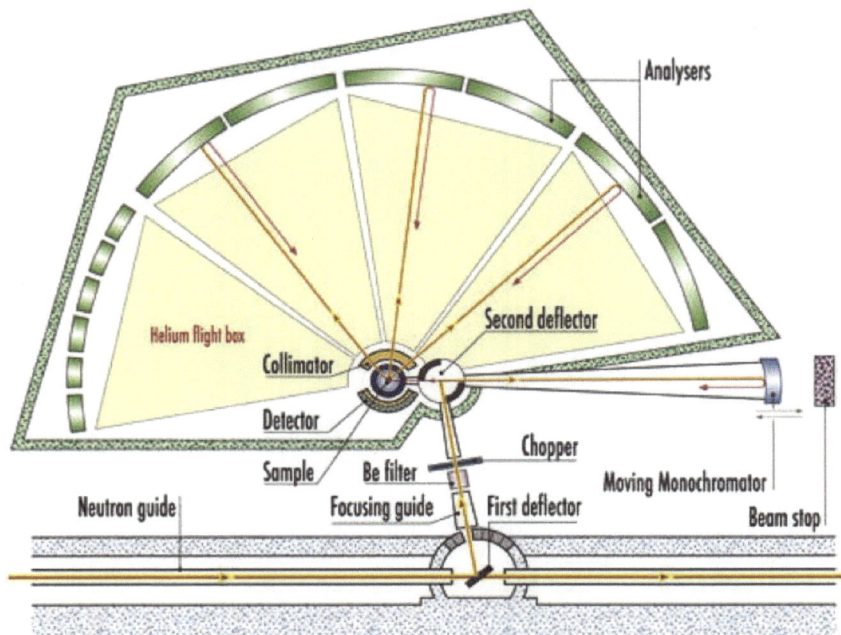

Figure 4: An example of an in-direct geometry instrument. IN16 at the ILL [45] uses a Doppler drive to generate ΔE_i. The figure, taken from [60], is courtesy of the ILL

Historically, the higher energy resolution (~ 1 μeV) backscattering instruments at reactor sources (*i.e.* IN16, ILL, France [45] or HFBS, NIST, USA [40]) use Bragg scattering from a single crystal to generate a highly monochromatic incident neutron beam, E_i. The crystal monochromator, however, is attached to a mechanical Doppler unit which, when oscillating, imparts a small Doppler shift (see Fig. **4**). The neutron beam is no longer highly monochromatic but exhibits a spread of incident energies, ΔE_i. While such instruments offer high energy resolution and access the nano-second time regime they do so at the expense of a sizable energy transfer window. Alternatively, backscattering instruments that employ the time of flight method use velocity-selectors (wide aperture disc choppers) to transmit a band of incident neutron energies. While not as sensitive as Doppler driven instruments, indirect t.o.f spectrometers do allow access to wide, tunable, dynamic ranges and can offer sizable improvements in incident neutron flux. Typically, such instruments offer similar flexibility to direct geometry instrumentation but straddle the pico-second to nano-second time range with an energy transfer resolution in the low μeV ($1 - 50$) regime.

Neutron Spin Echo (NSE) and QENS

Compared to backscattering techniques, neutron spin echo extends the accessible temporal and spatial ranges to longer times (nanoseconds) and larger length scales. Indeed, the spatial range covered by NSE and by small angle neutron scattering (SANS) partially overlap. As a result, should SANS be used to measure structural conformations *via* S(Q), NSE can be used as a complementary tool to probe associated dynamics. This unique technique was pioneered by F. Mezei [61] and is described in detail in [62]. The ability to directly observe a change in the velocity of a neutron during the scattering process is key to the high energy resolution (nano-electron volts) afforded by NSE. The technique allows measurement of 3 to 4 orders of magnitude in time. For a review of the NSE as applied to soft matter systems see [63].

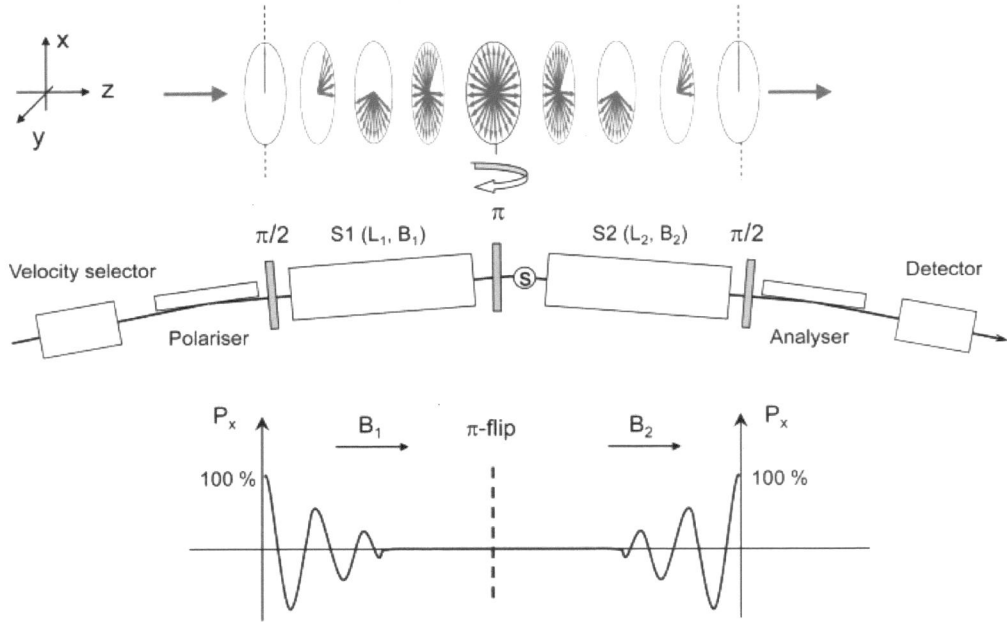

Figure 5: The spin echo method and basic spectrometer components

By following Fig. **5**, the technique can be illustrated as follows. A super-mirror assembly longitudinally spin polarises a broadly monochromatic beam of neutrons ($\Delta\lambda/\lambda \sim 15\ \%$). The polarised beam is rotated by $\pi/2$ radians into the x-direction before travelling along the (z-) axis of solenoid, S_1 (length = L_1, longitudinal field strength B_1). The neutrons enter S_1 with their spin polarisation perpendicular to the field direction and Larmor precess in the x-y plane. The polarisation component of the beam along the x direction perpendicular to B_1 is,

$$P_x = <\cos(\varphi)> = \int f(v)\cos\left(\frac{\gamma_L \int_L B_1 dl}{v}\right) dv \tag{12}$$

Here φ is the total precession angle (in radians) over the distance L. γ_L is the gyro-magnetic ratio of the neutron and f(v) is the velocity distribution function. Since the incident neutron beam is broadly monochromatic, the precessing spins associated with neutrons of differing velocities will de-phase and $P_x \rightarrow 0$ at the exit of the first solenoid, or arm. Before passing through the second solenoid (length = L_2, longitudinal field strength = B_2) the neutron spins are flipped through π radians. The de-phased neutron spins again precess in the x-y plane of S_2 *but* in the opposite sense to that observed in S_1. For elastic scattering the velocity distributions of the neutron beam before and after the sample are identical. At the 'echo' condition, when the solenoid field integrals are equal ($\int_{S_1} B_1 dl = \int_{S_2} B_2 dl$),

then the same number of spin precessions occur in both S_1 and S_2 and the total precessional angle is zero for each neutron in the beam. Thus, irresepective of $\Delta\lambda/\lambda$, full neutron spin polarisation is recovered at the exit of the second arm. The net spin polarisation of the transmitted beam is determined by flipping the neutrons back into the longitudinal (z-) direction (using a second $\pi/2$ flipper) and analysing their spin component using a second supermirror polariser. Considering for simplicity a perfectly monochromatic beam, the velocity distribution function is a delta function centered at $v = v_1$. If φ_1 and φ_2 are the final precession angles after the first and second arms respectively, then the total precession angle, φ_T, over a total flight path $L_T = L_1 + L_2$ is,

$$\varphi_T(v_1, v_2) = \varphi_1 - \varphi_2 = \gamma_L \left(\frac{\int_{S_1} B_1 dl}{v_1} - \frac{\int_{S_2} B_2 dl}{v_2} \right) \tag{13}$$

where v_2 is the velocity of the scattered neutron. For a quasi-elastic scattering event, the velocity of a scattered neutron will change and $v_2 = v_1 + \delta v_1$. For solenoid fields set to ensure the 'echo' condition, quasi-elastic scattering will result in an unequal number of neutron precessions in the two solenoids arms and thus loss of polarisation. The energy transfer associated with the QENS process is related to the velocity of the incident and scattered neutrons *via* $\Delta E = E_1 - E_2 = \frac{1}{2} m_n (v_2^2 - v_1^2) = \hbar \omega(v_1, v_2)$. As a result $\varphi_T(v_1, v_2)$, and hence spin polarisation, can be related to the energy transfer, $\hbar\omega(v_1, v_2)$. In the quasielastic limit $v_2 = v_1 + \delta v_1 = v_1 + \hbar\omega/m_n v_1$ and $\Delta E = \hbar\omega = m_n v_1 \delta v_1$. If $v_2 = v_1 + \hbar\omega/m_n v_1$ is incorporated into Eq. (13), and the result expanded to the first order in ω, the total accumulated precession angle, φ_T, becomes,

$$\varphi_T = \frac{\gamma_L \left(\int_{S_1} B_1 dl - \int_{S_2} B_2 dl \right)}{v_1} + \frac{\gamma_L \int_{S_2} B_2 dl}{m_n v_1^3} \hbar\omega \tag{14}$$

At the echo condition only the second term stands and the accumulated precession angle is,

$$\varphi_T' = \frac{\gamma_L \int_{S_2} B_2 dl}{m_n v_1^3} \hbar\omega = t_f \omega \tag{15}$$

t_f is a constant of proportionality which has the units of time. t_f is directly proportional to i) the field integrals and ii) λ^3 (λ is the incident wavelength). Therefore, for quasi-elastic scattering, the spin polarisation at the echo condition will decrease to a value of $\cos(\varphi_T')$, or $\cos(\omega t_f)$. The scattering function is a measure of the distribution of precession angles in the scattered beam since, as discussed in [23], it describes the probability that a neutron is scattered with an energy change, $\hbar\omega$. The net polarisation of the beam can therefore be averaged over all QENS processes to give,

$$P_z(\mathbf{Q}, t_f) = \langle \cos(\omega t_f) \rangle = \frac{\int S(\mathbf{Q}, \omega) \cos(\omega t_f) d\omega}{\int S(\mathbf{Q}, \omega) d\omega} = \frac{I(\mathbf{Q}, t_f)}{I(\mathbf{Q})} \tag{16}$$

The numerator represents the Fourier transform of $S(\mathbf{Q}, \omega)$ with respect to ω. The denominator is the integral of $S(\mathbf{Q}, \omega)$ over all energy *i.e.* the static structure factor $S(\mathbf{Q})$. A measurement of the final polarisation of the beam for a given t_f is simply a measurement of the normalised intermediate scattering law, $I(\mathbf{Q}, t_f) / I(\mathbf{Q})$ where, in the quasi-elastic limit, the Fourier time, t_f, is equivalent to real time. This general result can be easily extended to the case of a broadly monochromatic beam of incident neutrons, defined by a velocity distribution function, $f(v)$, as used for a real experiment.

EXPERIMENTAL CONSIDERATIONS

The measured scattering function, $S^{meas}_{inc}(Q, \omega)$, is the convolution of $S_{inc}(Q, \omega)$ and the resolution function of the neutron instrument, $R(Q, \omega)$. For spectrometers operating in Q-ω space,

$$S^{meas}_{inc}(\mathbf{Q}, \omega) = S_{inc}(\mathbf{Q}, \omega) \otimes R(\mathbf{Q}, \omega) \qquad (17)$$

In its simplest form the instrument resolution approximates to a Gaussian or Lorentzian function of finite width, Γ_{res} (usually quoted as full width at half maximum). The design of any neutron instrument attempts to minimise Γ_{res}, yet maintain a wide energy transfer window, by optimising the beam line neutronics (*i.e.* timing uncertainties, analyser crystal characteristics and geometry). It can be shown that the width of the resolution function affects an instrument's experimental observation time [64]; $\tau \propto 1/\Gamma_{res}$. As a result, any observed quasi-elastic broadening will be associated with motions, $\tau \cong 1/\Gamma_{res}$. Slower dynamics will be resolution limited (*i.e.* 'hidden' within the resolution function), while fast processes will result in an extremely broad, background-like, component. Ideally, several energy resolutions, and thus neutron spectrometers, should be used to avoid a resolution-biased view of any given problem. For a well defined resolution, R(Q,ω) can be generated using its functional form. In the case of more complex resolution functions, R(Q,ω) should be measured during an experiment. In general, this measurement is performed using a standard; typically a piece of vanadium similar in size and orientation to the sample. However, R(Q,ω) can be also determined from the sample itself if, for example, the sample is cooled to temperature where only elastic scattering is observed. The latter avoids the geometric uncertainties associated with a resolution function measured using a standard. Extracting $S_{inc}(Q,\omega)$ from an experimental dataset requires de-convolution of $S^{meas}_{inc}(Q,\omega)$ and R(Q,ω). Before de-convolution, however, the measured spectra are normalised to the number and/or wavelength distribution of the incident neutron beam and corrected for beam attenuation effects within the sample (*i.e.* absorption corrections). The signal from an empty sample container is also removed. Multiple scattering (MS) corrections may be applied [23] but in practice the sample thickness is optimised (*i.e.* ~ 10% scatterer) such that MS effects are deemed minimal. Finally, using either a measured or theoretical R(Q,ω), least squares fitting or Bayesian analysis [65, 66] routines are applied to the corrected dataset to isolate the intensities and widths of the spectral contributions to $S_{inc}(Q,\omega)$. It should be noted that care must be taken when interpreting a scattering function measured on a time-of-flight instrument. Since neutron detectors are located at a fixed scattering angles, θ, a measure of $S_{inc}(Q,\omega)$ on a t.o.f instrument is in fact a measure of $S_{inc}(\theta,\omega)$. Ideally, $S_{inc}(\theta,\omega)$ should be transformed to Q-ω space before spectral analysis. It should also be mentioned that for a NSE measurement, the instrument resolution is removed from a measured data set by simply dividing each data point by I(Q,t) collected from an elastically scattering sample. Theoretically, I(Q,t) could be obtained from S(Q,ω) measured using a backscattering instrument *via* its Fourier transform. In practice, however, experimental limitations, such as truncation errors, limit the efficacy of the results.

EXAMPLES

While it is not the purpose of this section to provide a comprehensive overview of all bio-macromolecular problems that have been addressed using quasi-elastic neutron spectroscopy (see [10, 11]), it is worth concluding with a few experimental results which highlight the ideas previously outlined.

    1.   Cytoplasmic water and hydration dynamics in human red blood cells (Stadler *et al.*, 2008 [67])

A detailed understanding of water - macromolecule interactions in cells is of major scientific interest. With macromolecular concentrations approaching 400 mgmL^{-1} [68], the cell is a crowded environment and distances between macromolecular species are typically 1 nm; the thickness of a few water layers. The behaviour of water that is in close contact with protein surfaces [69] (be they hydrophilic or hydrophobic) or trapped in surface cavities [70] is shown to deviate from that observed from bulk liquid. However, studies [71] show that in cells and bacteria a major fraction of H_2O exhibits bulk like dynamics.

Using quasi-elastic neutron scattering, and by combining data from three neutron spectrometers, IRIS (time scale ~ 40 ps), FOCUS (~ 13 ps) and TOFTOF (~ 7 ps), the dynamics of water in human red blood cells, *in vivo*, were investigated in the temperature range 290 and 320 K see Fig. **6(a-d)**. Red blood cells (RBC) consist mostly of haemoglobin and water at a concentration 330 mgmL^{-1} [72]. A combination of neutron instruments was necessary to span the time scales associated with bulk water dynamics and reduced mobility interfacial water motions. The RBCs were suspended in an aqueous buffer and concentrated by centrifugation. Residual extracellular water content was found to be less than 10 % of the total water content. Deuteration techniques were used to isolate the cytoplasmic dynamics of H_2O from membrane and macromolecular dynamics. This was done by subtracting experimental QENS data of natural abundance red blood cells in a D_2O buffer from natural abundance RBC in a H_2O buffer. In addition, QENS measurements were collected from a pure H_2O buffer as a reference.

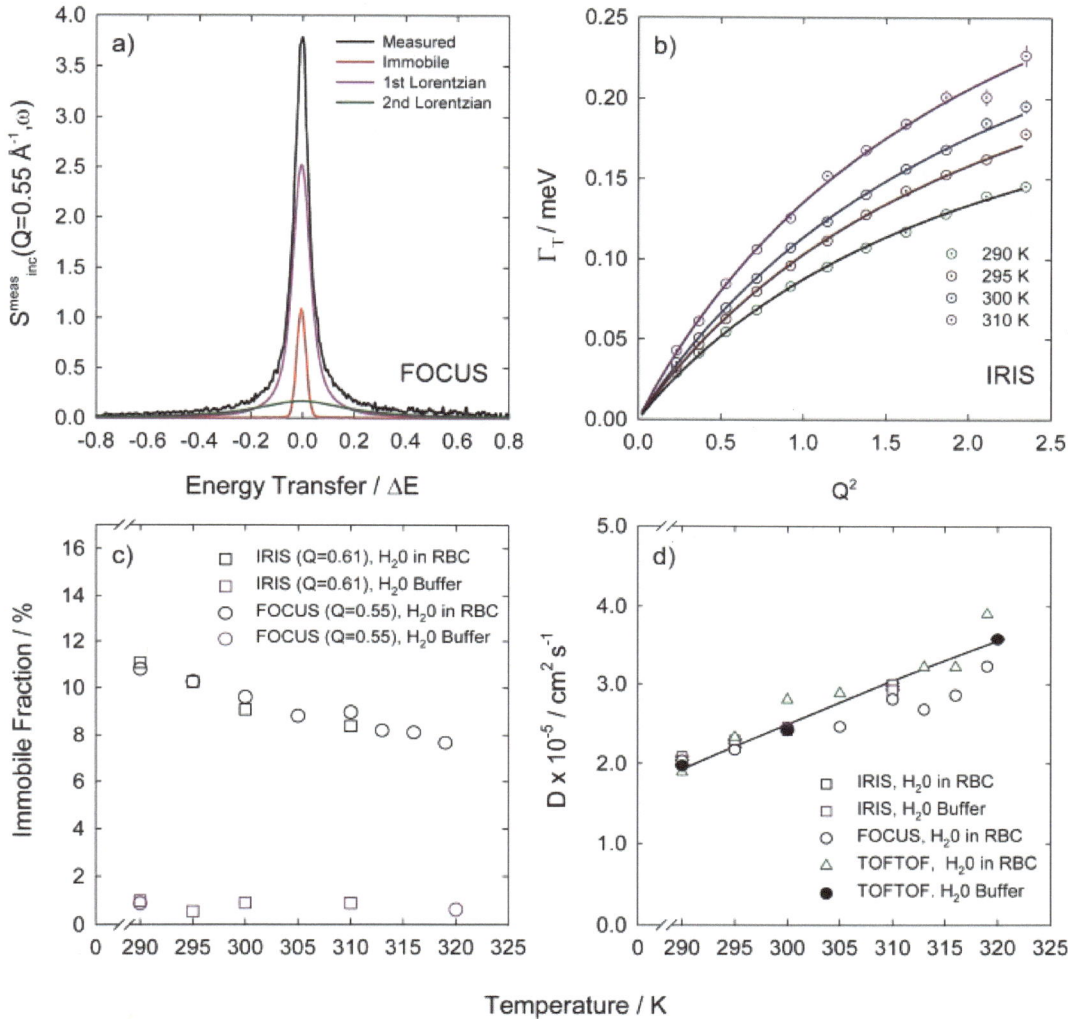

Figure 6: (a) A typical QENS spectrum collected using FOCUS (290 K, Q = 0.55 Å^{-1}) and the spectral components associated with cytoplasmic H_2O dynamics in RBC. The immobile fraction was determined from the amplitude of the elastic component. b) The linewidths (half width at half-maximum) associated with the translational diffusion process of cytoplasmic H_2O in RBC. The solid lines represent fits with the jump-diffusion model of Singwi and Sjölander. c) The temperature dependence of the immobile fraction (%) of cytoplasmic H_2O in RBC as measured on IRIS and FOCUS (Q = 0.61 and 0.55 Å^{-1} respectively). Such small scattering vectors allow access to length scales ~ 10 Å which enable the confining effect on water by protein surface cavities and boundaries to be observed. The immobile fraction (%) of the H_2O buffer is shown for comparison. d) Translational diffusion coefficient, *D,* of cytoplasmic H_2O in RBC measured on IRIS, FOCUS and TOFTOF. The translational diffusion coefficient from the H_2O buffer, as measured on IRIS and TOFTOF, is also shown. The solid line is the normal temperature dependence of an H_2O buffer.

Using the QENS technique, it has been shown that a major fraction (~90 %) of cell water can be characterized by a translational diffusion coefficient similar to that of bulk water; a result in agreement with results observed from QENS [71] and NMR [73] studies of *Escherichia coli*. The remaining fraction (~10%) of cellular water, however, exhibits reduced, or slow, dynamics and is attributed to dynamically bound water on the surface of haemoglobin which accounts for approximately half of the hydration layer. The properties of this bound hydration water should have a discernable influence on protein stability and interaction in living cells.

2.   A benchmark for protein dynamics: Ribonuclease A measured by neutron scattering in a large wave-vector energy transfer range (Wood *et al.*, 2008 [74])

Ribonuclease A (RNase A) is a 124 residue, single domain protein which catalyses the cleavage of RNA. Historically one of the first enzymes to be isolated, and originally used in the study of protein folding, work on this material continues in several fields [75-77]. More recently, however, RNase A has been used as marker for the detailed experimental characterisation of protein dynamics (internal motions and whole molecule diffusion characteristics) arising from stabilisation forces. By combining time-of-flight / backscattering and spin echo instrumentation, the aim of this study was to amass a detailed understanding of the dynamic landscape of the RNase A enzyme over a wide temporal and spatial range. The results would act as reference for neutron studies of other proteinaceous materials.

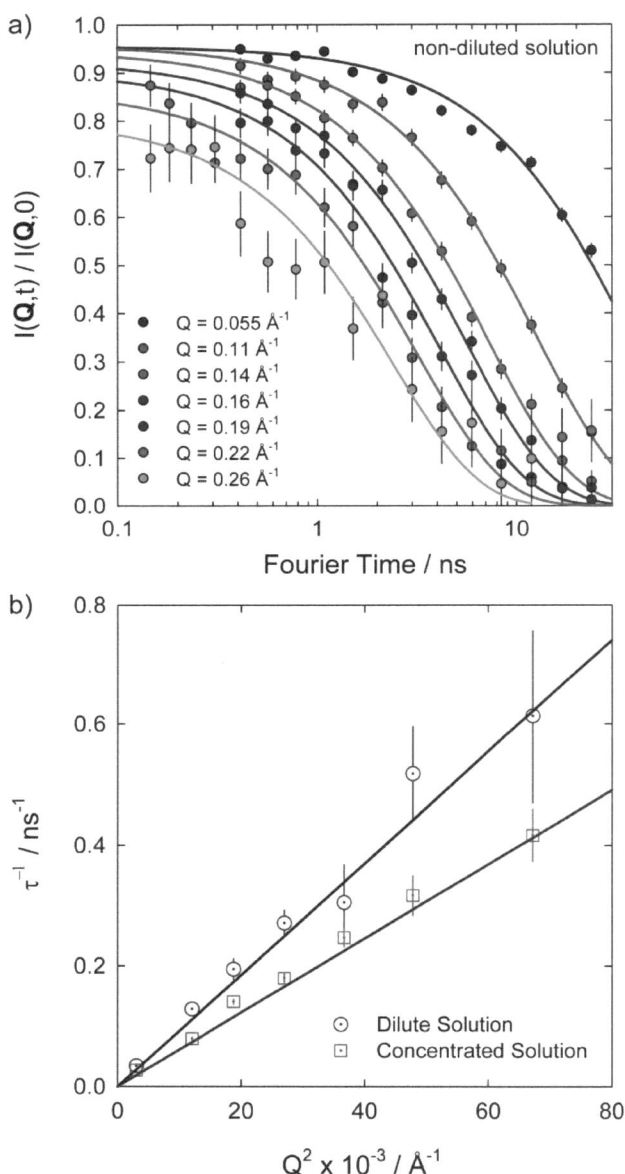

Figure 7: a) Normalised I(Q,t) from the concentrated protein solution measured at room temperature. All data sets are fitted using a single exponential relaxation function b) Relaxation times determined from the dilute and concentrated protein solutions. The lines are linear fits to the data and give diffusion coefficients of $D_{dil} = 1.03 \times 10^{-6}$ cm^2 s^{-1} and $D_{conc} = 0.67 \times 10^{-6}$ cm^2 s^{-1}.

Internal dynamics were probed by measuring hydrated (0.4 gram D_2O per gram protein) and dehydrated powders on direct geometry time-of-flight and non-t.o.f backscattering neutron instruments. In contrast, whole molecule, or global, protein diffusion parameters were ascertained from protein solutions at the level of 200 mgmL^{-1} D_2O buffer (concentrated solution) and 30 mgmL^{-1} (dilute solution) using the neutron spin echo technique. The internal dynamic processes observed in RNase A are summarised in [74]. Here, we highlight the applicability of the neutron spin echo technique for the study of biological macromolecules by focussing on the extraction of the global protein diffusion coefficients.

Whole molecule diffusion was characterised using the IN11 NSE instrument at the Institut Laue-Langevin. Using an incident neutron wavelength of 7.8 Å, the study was performed over a Q range 0.06 – 0.27 Å^{-1} (associated length scales ($2\pi/Q$) ~ 20 to 100 Å) with an energy resolution of 300 neV. The experimental Fourier time window ranged from 26 ps to 24 ns.

Normalised I(Q,t) data from both samples was well described using a single exponential relaxation function (Fig. **7(a)**). The relaxation time, τ, was seen to increase with increasing momentum transfer. Although, the experimental time window afforded by IN11 was not sufficient to completely detect the relaxation process at small Q-values, complete relaxation was observed by 20 ns for higher wave vectors. The single exponential description of the data and the linear dependence of τ with Q^2 indicated that the relaxation observed was associated with center-of-mass diffusion with the diffusion coefficient, D, being given by $\tau^{-1} = DQ^2$. Slower diffusive behaviour, however, was observed from the concentrated sample. For the dilute system the experimentally determined value for D_{dil} was found to be comparable to that predicted for RNase A (in solution) [78] using MD simulations.

    3.   An-harmonic behaviour in the multi-subunit protein, apoferritin (Telling *et al.*, 2008 [79])

From a neutron perspective, *via* measurement of the mean squared displacement (msd) parameter, the effect of hydration on a protein's dynamic landscape can exhibit itself *via* the so-called dynamical transition temperature, T_d [80-82]. While care should be taken interpreting T_d [6] it is, in broad terms, a temperature indicative of additional degrees of conformational freedom above which the material can no longer be considered as a vibrating harmonic solid. Typically occurring above 200 K, T_d is not seen in lyophilized material.

Apoferritin is the Fe depleted form of ferritin, the natural iron storage protein. The apoferritin molecule can be thought of as a multi-subunit spherical shell of internal diameter ~ 8 nm. This shell, which is ~ 2 nm thick, is composed of 24 polypeptide chains [83]. While apoferritin exhibits a 'dynamic transition' in the hydrated state (h ≥ 0.14 gram water per gram protein) a weaker hydration-independent inflection in the msd parameter is observed at lower temperatures (~100K) see Fig. **8** (top). Analysis of elastic neutron scattering measurements collected from lyophilised material using the IN16 and OSIRIS spectrometers see Fig. **8** (bottom) suggest that the mechanism responsible for the hydration-independent inflection at ~ 100 K is the re-orientation, or activation, of methyl groups (CH_3). Methyl groups contribute ~ 25 – 30 % of all non-exchangeable protons in the protein. In fact, analysis using models developed to describe side group motion in glassy polymers, namely,

$$S(Q,\omega \approx 0, T) = DWF \times \left(A_o(Q) + \frac{2}{\pi}[1 - A_o(Q)] \times \sum g_i \arctan\left[\frac{\Gamma_{res}}{\Gamma}\right] \right) \tag{18}$$

suggests that the CH_3 groups exhibit a broad distribution of activation energies. Here g_i gives the weight of each component according to a Gaussian distribution of activation energies, Γ_{res} is the width of the instrument resolution function (fwhm) and Γ is the width of the Lorentzian line characterising the quasi-elastic broadening. $A_o(Q)$ ($= 1/3 * [1 + 2 j_o(\sqrt{3}Qr)]$) is the EISF describing a 3-site jump re-orientation. j_o is a zero-order Bessel function and r (= 1.032 Å [84]) is the distance between moving protons. Similar results are reported from other proteinaceous materials in which a high percentage of non-exchangeable protons associated with CH_3 [85, 86]. In contrast, with less than 5% of all non-exchangeable protons are associated with methyl species, RNA [87], shows no such inflexion.

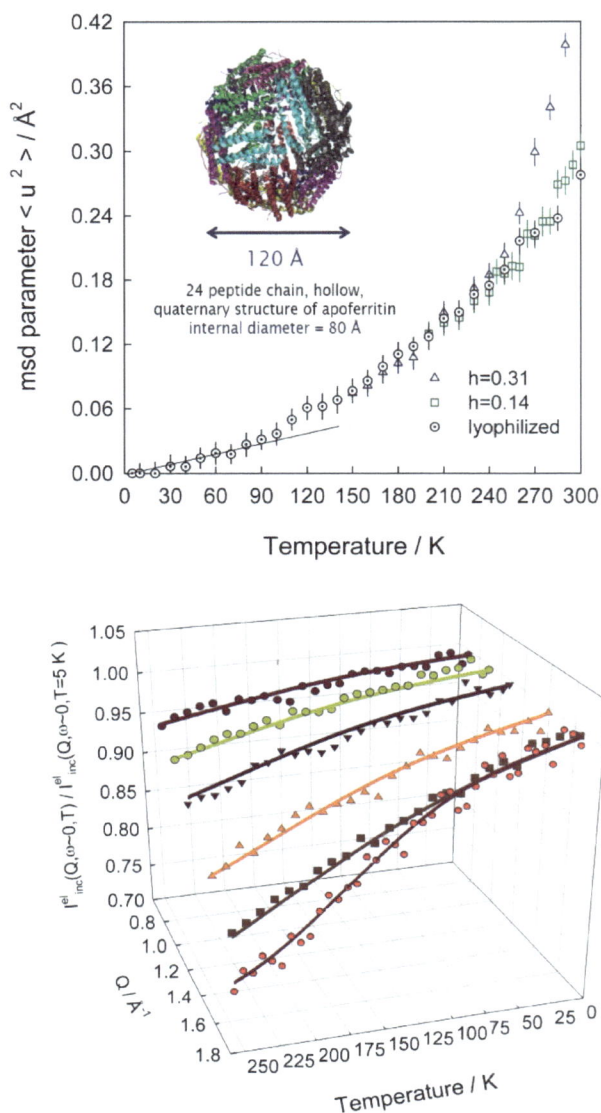

Figure 8: Top - The hydration dependence of the msd parameter $<u^2>$ as a function of temperature for lyophilized, h = 0.14 and h = 0.31 apoferritin as seen on OSIRIS. Inset: the apoferritin quaternary structure. Bottom - Elastic window scans from lyophilized apoferritin at Q = 0.744 (●), 0.94 (●), 1.123 (▼), 1.443 (▲) and 1.684 Å^{-1} (■) ; a Q-range straddling length scales from 3.7 to 8.5 Å. The solid lines are the result of simultaneously fitting Eq. (18) to the data. Data, and subsequent fit to Eq. (18), from the IN16 is also shown for Q = 1.66 Å^{-1} (●).

ACKNOWLEDGMENTS

This section could not have been completed without generous input from: P. Fouquet, B. Frick, T. Seydel, J. Ollivier, F. Natali and A. Stadler (ILL), K. Wood (ANSTO), A. Faraone and M. Nagao (NIST), F. Juranyi (PSI), J. Wuttke and O. Holderer (JCNS), G. Ehlers and E. Mamontov (ORNL), T. Unruh (FRM), K. Nakajima (J-PARC), R.E. Lechner (HMI) and V. García Sakai (ISIS). MTFT would also like to thank Dr B. Gabrys and Prof. C. Grovenor (Department of Materials, University of Oxford, UK).

REFERENCES

[1]   https://www.ill.eu/sites/deuteration/index.htm

[2]   http://www.ansto.gov.au/research/bragg_institute/facilities/molecular_deuteration/

[3]   Pieper J, Hauss T, Buchsteiner A, Renger G. The effect of hydration on protein flexibility in photosystem II of green plants studied by quasielastic neutron scattering. Eur Biophys J 2008; 37: 657-663.

[4]   http://neutron.neutron-eu.net/n_ess

[5]   http://xfel.desy.de/

[6]   Doster W. The dynamical transition of proteins, concepts and misconceptions. Eur Biophys J 2008; 37: 591-602.

[7]   Wood K, Plazanet M, Gabel F, Kessler B, Oesterhel D, Tobias DJ, Zaccai G, Weik M. Coupling of protein and hydration-water dynamics in biological membranes. Proc Nat Acad Sci USA 2007; 104: 18049-18054.

[8]   Flenner E, Das J, Rheinstadter MC, Kosztin I. Subdiffusion and lateral diffusion coefficient of lipid atoms and molecules in phospholipid bilayers. Phys Rev E 2009; 79: 1-11.

[9]   Glass DC, Krishnan M, Nutt DR, Smith JC. Temperature Dependence of Protein Dynamics Simulated with Three Different Water Models. J Chem Theory Comput 2010; 6: 1390-1400.

[10]   Fitter J, Gutberlet T, Katsaras J, (Eds.). Neutron Scattering in Biology: Techniques and Applications. Berlin, Springer, 2006.

[11]   Sakai VG, Arbe A. Quasielastic neutron scattering in soft matter. Curr Opin Coll Interf Sci 2009; 14: 381-390.

[12]   Gabel F, Bicout D, Lehnert U, Tehei M, Weik M, Zaccai G. Protein dynamics studied by neutron scattering. Q Rev Biophys 2002; 35: 327-367.

[13]   Rheinstaedter MC, Seydel T, Haeubler W, Salditt T. The 'neutron window' of collective excitations in lipid membranes. Physica B-Cond Mat 2006; 385: 722-724.

[14]   Ferrand M, Dianoux AJ, Petry W, Zaccai G. Thermal Motions and Function of Bacteriorhodopsin in Purple Membranes - Effects of Temperature and Hydration Studied by Neutron-Scattering. Proc Nat Acad Sci USA 1993; 90: 9668-9672.

[15]   Swenson J, Kargl F, Berntsen P, Svanberg C. Solvent and lipid dynamics of hydrated lipid bilayers by incoherent quasielastic neutron scattering. J Chem Phys 2008; 129: 045101.

[16]   Doxastakis M, Sakai VG, Ohtake S, Maranas JK, de Pablo JJ. A molecular view of melting in anhydrous phospholipidic membranes. Biophys J 2007; 92: 147-161.

[17]   Roh JH, Briber RM, Damjanovic A, Thirumalai D, Woodson SA, Sokolov AP. Dynamics of tRNA at Different Levels of Hydration. Biophys J 2009; 96: 2755-2762.

[18]   Telling M. A sweeter understanding of cryo-preservation Mat Today 2009; 12: 68-68.

[19]   Magazu S, Migliardo F, Telling MTF. Structural and dynamical properties of water in sugar mixtures. Food Chem 2008; 106: 1460-1466.

[20]   Varga B, Migliardo F, Takacs E, Vertessy B, Magazu S, Telling MTF. Study of solvent-protein coupling effects by neutron scattering. J Biol Phys 2010; 36: 207-220.

[21]   Ghugare SV, Mozetic P, Paradossi G. Temperature-Sensitive Poly(vinyl alcohol)/Poly(methacrylate-co-N-isopropyl acrylamide) Microgels for Doxorubicin Delivery. Biomacromol 2009; 10: 1589-1596.

[22]   Telling M. Magic bullets and plastic sponges - developing novel drug delivery materials Mat Today 2009; 12: 65-65.

[23]   Bée M. Quasi-Elastic Neutron Scattering: Principles and Applications in Solid State Chemisty, Biology and Materials Science. Bristol, Adam Hilger, 1988.

[24]   Higgins JS, Benoit HC. Polymers and Neutron Scattering. Oxford, Clarendon Press, 1996.

[25]   Windsor CG. Pulsed Neutron Scattering. London, Taylor and Francis, 1981.

[26]   Sears VF. Neutron scattering lengths and cross sections. Neutron News 1992; 3: 26-37.

[27]   Van Hove L. Correlations in Space and Time and Born Approximation Scattering in Systems of Interacting Particles. Phys Rev 1954; 95: 249-262.

[28]   Teixeira SCM, Zaccai G, Ankner J, Bellissent-Funel MC, Bewley R, Blakeley MP, Callow P, Coates L, Dahint R, Dalgliesh R, Dencher NA, Forsyth VT, Fragneto G, Frick B, Gilles R, Gutberlet T, Haertlein M, Hauss T, Haussler W, Heller WT, Herwig K, Holderer O, Juranyi F, Kampmann R, Knott R, Krueger S, Langan P, Lechner RE, Lynn G, Majkrzak C, May RP, Meilleur F, Mo Y, Mortensen K, Myles DAA, Natali F, Neylon C, Niimura N, Ollivier J, Ostermann A, Peters J, Pieper J, Ruhm A, Schwahn D, Shibata K, Soper AK, Strassle T, Suzuki J, Tanaka I, Tehei M, Timmins P, Torikai N, Unruh T, Urban V, Vavrin R, Weiss K. New sources and instrumentation for neutrons in biology. Chem Phys 2008; 351: 170-170.

[29]   Lechner RE. NEAT experiments at BENSC. Neutron News 1996; 7: 9-11.

[30]   Unruh T, Neuhaus E, Petry W. The high-resolution time-of-flight spectrometer TOF-TOF Nuclear Instruments & Methods in Physics Research Section a-Accelerators Spectrometers Detectors and Associated Equipment 2008; 585: 201-201.

[31]   Ollivier J, Mutka H, Didier L. The New Cold Neutron Time-of-Flight Spectrometer IN5. Neutron News 2010; 21: 22 - 25.

[32]   http://www.ill.eu/instruments-support/instruments-groups/instruments/in6/

[33]   Janssen S, Mesot J, Holitzner L, Furrer A, Hempelmann R. FOCUS: A hybrid TOF-spectrometer at SINQ. Physica B 1997; 234: 1174-1176.

[34]   Copley JRD, Cook JC. The Disk Chopper Spectrometer at NIST: a new instrument for quasielastic neutron scattering studies. Chem Phys 2003; 292: 477-485.

[35]   http://www-llb.cea.fr/spectros/pdf/mibemol-llb.pdf

[36]   Nakajima K, Nakamura M, Kajimoto R, Osakabe T, Kakurai K, Matsuda M, Metoki M, Wakimoto S, Sato TJ, Itoh S, Arai M, Yoshida K, Niita K. Cold-neutron disk-chopper spectrometer at J-PARC. J Neutron Res 2007; 15: 13 - 21.

[37]   http://www.ansto.gov.au/research/bragg_institute/facilities/instruments/pelican

[38]   http://www.isis.stfc.ac.uk/instruments/let/

[39]   Mason TE, Abernathy D, Anderson I, Ankner J, Egami T, Ehlers G, Ekkebus A, Granroth G, Hagen M, Herwig K, Hodges J, Hoffmann C, Horak C, Horton L, Klose F, Larese J, Mesecar A, Myles D, Neuefeind J, Ohl M, Tulk C, Wang XL, Zhao J. The Spallation Neutron Source in Oak Ridge: A powerful tool for materials research. Physica B-Cond Mat 2006; 385-86: 955-960.

[40]   Meyer A, Dimeo RM, Gehring PM, Neumann DA. The High Flux Backscattering Spectrometer at the NIST Center for Neutron Research. Rev Sci Instrum 2003; 74: 2759.

[41]   Telling MTF, Andersen KH. Spectroscopic characteristics of the OSIRIS near-backscattering crystal analyser spectrometer on the ISIS pulsed neutron source. Physical Chemistry Chem Phys 2005; 7: 1255-1261.

[42]   Carlile CJ, Adams MA. The Design of the Iris Inelastic Neutron Spectrometer and Improvements to Its Analyzers. Physica B 1992; 182: 431-440.

[43]   Mamontov E, Zamponi M, Hammons S, Keener WS, Hagen M, Herwig KW. BASIS: A New Backscattering Spectrometer at the SNS. Neutron News 2008; 19: 22 - 24.

[44]   Randl OG, Franz H, Gerstendorfer T, Petry W, Vogl G, Magerl A. How to rejuvenate an old lady: New crystals for the backscattering spectrometer IN10. Physica B 1997; 234: 1064-1065.

[45]   Frick B, Gonzalez M. Five years operation of the second generation backscattering spectrometer IN16 - a retrospective, recent developments and plans. Physica B-Cond Mat 2001; 301: 8-19.

[46]   Frick B, Bordallo HN, Seydel T, Barthelemy JF, Thomas M, Bazzoli D, Schober H. How IN16 can maintain a world-leading position in neutron backscattering spectrometry. Physica B-Cond Mat 2006; 385-86: 1101-1103.

[47]   Frick B, Mamontov E, van Eijck L, Seydel T. Recent Backscattering Instrument Developments at the ILL and SNS. Zeitschrift Fur Physikalische Chemie-Int J Res Phys Chem Chem Phys 2010; 224: 33-60.

[48]   Francesca N, Peters J, Russo D, Barbieri S, Chiapponi C, Cupane A, Deriu A, Di Bari MT, Farhi E, Gerelli Y, Mariani P, Paciaroni A, Rivasseau C, Schirò G, Sonvico F. IN13 Backscattering Spectrometer at ILL: Looking for Motions in Biological Macromolecules and Organisms. Neutron News 2008; 19: 14-18.

[49]   Tregenna-Piggott PLW, Juranyi F, Allenspach P. Introducing the inverted-geometry time-of-flight backscattering instrument, MARS at SINQ. J Neutron Res 2008; 16: 1-12.

[50]   http://www.jcns.info/jcns_spheres/

[51]   Farago B. IN11C, medium-resolution multidetector extension of the IN11 NSE spectrometer at the ILL. Physica B 1997; 241: 113-116.

[52]   Farago B. Recent neutron spin-echo developments at the ILL (IN11 and IN15). Physica B-Cond Mat 1999; 267: 270-276.

[53]   Schleger P, Ehlers G, Kollmar A, Alefeld B, Barthelemy JF, Casalta H, Farago B, Giraud P, Hayes C, Lartigue C, Mezei F, Richter D. The sub-neV resolution NSE spectrometer IN15 at the Institute Laue-Langevin. Physica B 1999; 266: 49-55.

[54]   Rosov N, Rathgeber S, Monkenbusch M. Neutron Spin Echo spectroscopy at the NIST Center for Neutron Research, in: Cebe P, Hsiao BS, Lohse DJ, (Eds.), Scattering from Polymers - Characterization by X-Rays, Neutrons, and Light, Washington, American Chemical Society, 2000.

[55]   Holderer O, Monkenbusch M, Schatzler R, Kleines H, Westerhausen W, Richter D. The JCNS neutron spin-echo spectrometer J-NSE at the FRM II. Meas Sci Tech 2008; 19: 1-7.

[56]   Häussler W, Gohla-Neudecker B, Schwikowski R, Streibl D, Böni P. RESEDA--The new resonance spin echo spectrometer using cold neutrons at the FRM-II. Physica B: Condensed Matter 2007; 397: 112-114.

[57]   Nagao M, Yamada NL, Kawabata Y, Seto H, Yoshizawa H, Takeda T. Relocation and upgrade of neutron spin echo spectrometer, iNSE. Physica B: Cond Mat 2006; 385-386: 1118-1121.

[58]   Fouquet P, Ehlers G, Farago B, Pappas C, Mezei F. The wide-angle neutron spin echo spectrometer project WASP. J Neutron Res 2007; 15: 39 - 47.

[59]   http://www.ill.eu/instruments-support/instruments-groups/instruments/in5/

[60]   http://www.ill.eu/instruments-support/instruments-groups/instruments/in16/

[61]   Mezei F. Neutron Spin-Echo - New Concept in Polarized Thermal-Neutron Techniques. Zeitsc Physik 1972; 255: 146-150.

[62]   Mezei F, Pappas C, Gutberlet T. Neutron Spin Echo. Springer, Berlin, 2003.

[63]   Farago B. Recent developments and applications of NSE in soft matter. Curr Opin Coll Interf Sci 2009; 14: 391-395.

[64]   Newport RJ, Rainford BD, Cywinski R. Neutron Scattering at a Pulsed Source. Bristol, Adam Hilger, 1988.

[65]   Sivia DS, Carlile CJ. Molecular-Spectroscopy and Bayesian Spectral-Analysis - How Many Lines Are There. J Chem Phys 1992; 96: 170-178.

[66]   Sivia DS, Carlile CJ, Howells WS, Konig S. Bayesian-Analysis of Quasi-Elastic Neutron-Scattering Data. Physica B 1992; 182: 341-348.

[67]   Stadler AM, Embs JP, Digel I, Artmann GM, Unruh T, Buldt G, Zaccai G. Cytoplasmic Water and Hydration Layer Dynamics in Human Red Blood Cells. J Am Chem Soc 2008; 130: 16852-16853.

[68]   Ellis RJ, Minton AP. Cell biology - Join the crowd. Nature 2003; 425: 27-28.

[69]   Russo D, Murarka RK, Copley JRD, Head-Gordon T. Molecular view of water dynamics near model peptides. J Phys Chem B 2005; 109: 12966-12975.

[70]   Makarov VA, Andrews BK, Smith PE, Pettitt BM. Residence times of water molecules in the hydration sites of myoglobin. Biophys J 2000; 79: 2966-2974.

[71]   Jasnin M, Moulin M, Haertlein M, Zaccai G, Tehei M. Down to atomic-scale intracellular water dynamics. Embo Reports 2008; 9: 543-547.

[72]   Krueger S, Nossal R. Sans Studies of Interacting Hemoglobin in Intact Erythrocytes. Biophys J 1988; 53: 97-105.

[73]   Persson E, Halle B. Cell water dynamics on multiple time scales. Proc Nat Acad Sci USA 2008; 105: 6266-6271.

[74]   Wood K, Caronna C, Fouquet P, Haussler W, Natali F, Ollivier J, Orecchini A, Plazanet M, Zaccai G. A benchmark for protein dynamics: Ribonuclease A measured by neutron scattering in a large wavevector-energy transfer range. Chem Phys 2008; 345: 305-314.

[75]   Dyer KD, Rosenberg HF. The RNase a superfamily: Generation of diversity and innate host defense. Mol Divers 2006; 10: 585-597.

[76]   Libonati M, Gotte G. Oligomerization of bovine ribonuclease A: structural and functional features of its multimers. Biochem J 2004; 380: 311-327.

[77]   Ribo M, Font J, Benito A, Torrent J, Lange R, Vilanova M. Pressure as a tool to study protein-unfolding/refolding processes: The case of ribonuclease A. Biochim Biophys Acta-Prot Proteom 2006; 1764: 461-469.

[78]   Tarek M, Tobias DJ. Subnanosecond Dynamics of Proteins in Solution: MD Simulations and Inelastic Neutron Scattering, in: Fitter J, Gutberlet T, Katsaras J, (Eds.), Neutron Scattering in Biology: Techniques and Applications, Berlin, Springer, 2006.

[79]   Telling MTF, Neylon C, Kilcoyne SH, Arrighi V. Anharmonic behavior in the multisubunit protein apoferritin as revealed by quasi-elastic neutron scattering. J Phys Chem B 2008; 112: 10873-10878.

[80]   Doster W, Cusack S, Petry W. Dynamical Transition of Myoglobin Revealed by Inelastic Neutron-Scattering. Nature 1989; 337: 754-756.

[81]   Filabozzi A, Deriu A, Andreani C. Temperature dependence of the dynamics of superoxide dismutase by quasi-elastic neutron scattering. Physica B 1996; 226: 56-60.

[82]   Wanderlingh UN, Corsaro C, Hayward RL, Bee M, Middendorf HD. Proton mobilities in crambin and glutathione S-transferase. Chem Phys 2003; 292: 445-450.

[83]   Donlin MJ, Frey RF, Putnam C, Proctor JK, Bashkin JK. Analysis of iron in ferritin, the iron-storage protein - A general chemistry experiment. J Chem Ed 1998; 75: 437-441.

[84]   Zhang CH, Arrighi V, Gagliardi S, McEwen IJ, Tanchawanich J, Telling MTF, Zanotti JM. Quasielastic neutron scattering measurements of fast process and methyl group dynamics in glassy poly(vinyl acetate). Chem Phys 2006; 328: 53-63.

[85]   Roh JH, Novikov VN, Gregory RB, Curtis JE, Chowdhuri Z, Sokolov AP. Onsets of anharmonicity in protein dynamics. Phys Rev Let 2005; 95: 1-4.

[86]   Doster W, Settles M. Protein-water displacement distributions. Biochim Biophys Acta-Prot Proteom 2005; 1749: 173-186.

[87]   Caliskan G, Briber RM, Thirumalai D, Garcia-Sakai V, Woodson SA, Sokolov AP. Dynamic transition in tRNA is solvent induced. J Am Chem Soc 2006; 128: 32-33.

CHAPTER 2

Elastic Incoherent Neutron Scattering: Biomolecular Motion Characterization by Self-Distribution-Function Procedure

Salvatore Magazù[1,*], Federica Migliardo[1,2] Antonio Benedetto[1], Miguel A. Gonzalez[3] and Claudia Mondelli[4]

[1]*Dipartimento di Fisica, Università di Messina, Viale D'Alcontres 31, S. Agata, P.O. Box 55, 98166 Messina Italy;* [2]*Laboratoire de Dynamique et Structure des Matériaux Moléculaires, UNESCO-L'Oréal University of Lille I, UMR CNRS 8024-59655 Villeneuve d'Ascq CEDEX, France ;* [3]*Institut Laue Langevin, 6, Rue Jules Horowitz, F-38042 Grenoble Cedex 9, France and* [4]*CNR-INFM-OGG and CRS Soft, Institut Laue Langevin, 6, Rue Jules Horowitz, F-38042 Grenoble Cedex 9, France*

Abstract: We first focus on the role of the instrumental resolution in Elastic Incoherent Neutron Scattering (EINS) where the connection between the Self Distribution Function (SDF) and the measured EINS intensity profile is highlighted. Second we show how the SDF procedure allows both the total and the partial Mean Square Displacement (MSD) evaluation, through the total and the partial SDFs. Finally, we compare the SDF and the Gaussian procedures, by applying the two approaches to EINS data collected, by the IN13 backscattering spectrometer (ILL, Grenoble), on aqueous mixtures of two homologous disaccharides, *i.e.* sucrose and trehalose, and on myoglobin.

INTRODUCTION

It is well known that the characterization of the different molecular processes involved in the dynamics of some molecular and macromolecular systems of biophysical interest, such as for example bioprotectant/water mixtures, pure and hydrated polymeric systems, hydrated and crystalline proteins, can be effectively investigated by evaluating the Mean Square Displacement (MSD) [1] from Elastic Incoherent Neutron Scattering (EINS) data collected by varying temperature, energy resolution, wave-vector and energy range and by using isotopic labelling. More specifically, a wide temperature range can facilitate the spectral separation of different molecular processes according to their time-scale, while the temperature dependence of the measured elastic intensity can provide information about the involved activation energies and, thus, the local potentials. On the other hand, a wide Q-range (e.g. such as that of IN13 backscattering spectrometer at the Institute Laue-Langevin (ILL) in Grenoble with a Q range extending up to 5 Å^{-1}) can allow to achieve a molecular assignment based on spatial features. Furthermore the use of proper energy window and energy resolution can allow to identify specific molecular motions and, as shown by Doster [2-4] with the so called "elastic resolution spectroscopy", can allow to derive the intermediate scattering function in the time domain from experiments performed with a different energy resolution. Finally the isotopic substitution, highlighting the contributions from specific system constituents, can allow to identify specific motions.

In the framework of the Gaussian approach the MSD can be obtained by a linear regression in the Guinier plot (where the logarithm of the elastic intensity is plotted as a function of Q^2) for a set of points close to Q=0. However this approach does not allow to separate the different contributions related to a specific spatial domain and furnishes MSD values which are dependent on the Q-range used for the MSD evaluation.

The aim of the present chapter is to clarify various aspects of the Self Distribution Function (SDF) procedure, proposed in different works [5-10] for evaluating the total and the partial MSDs, and to discuss the role of the instrumental resolution function in extracting the MSD. On that score we apply the procedure to EINS data collected by using the IN13 spectrometer at ILL on aqueous mixtures of two homologous disaccharides (*i.e.* sucrose and trehalose) and on dry myoglobin in trehalose environment.

Hydrated disaccharides are nowadays the object of intense research efforts motivated both by fundamental research and by their biotechnological applications. In particular, among disaccharides, trehalose has received a growing

Address correspondence to Salvatore Magazù: Dipartimento di Fisica, Università di Messina, P.O. Box 55, I-98166 Messina, Italy; E-mail: smagazu@unime.it

attention, because of both its wide role in nature and its potential use as a highly efficient natural bioprotectant system. Trehalose and sucrose aqueous mixtures have been characterized by light scattering, e.g. Photon Correlation Spectroscopy and Raman scattering, by neutron scattering, *i.e.* neutron diffraction, inelastic scattering and Quasi Elastic Neutron Scattering (QENS), and by simulation studies [11-14]. These techniques pointed out that trehalose shows a higher solute-solvent interaction strength, a higher kosmotropic character and a higher capability of dynamics switching off than sucrose.

Myoglobin stands out as the first structurally determined protein, and has been the subject of many detailed studies by a large number of experimental and computational methods [15-20]. Cordone and co-workers [21,22] have shown that the mean square displacements and the density of state function are those of a harmonic solid, up to room temperature and that the amplitude of the no-harmonic motions stemming from the inter-conversion among the protein's conformational substates is reduced with respect to the H_2O-solvated system, while their onset is shifted toward higher temperature.

It is well known that the scattering law $S(Q,\omega)$ is connected, in Planck's units, through its time Fourier transform (F_t), to the intermediate scattering function $I(Q,t)$, and, through its space-time Fourier transform ($F_{r,t}$), to the time-dependent spatial correlation functions $G(r,t)$ [23,24]. The scattering law $S(Q,\omega)$ is proportional to the observed neutron intensity, the proportionality factors being represented by the incident and outgoing neutron wave vectors, the number of scattered atoms and the scattering cross-sections (e.g., σ_{inc}(hydrogen)=81,0 barn and σ_{inc}(deuterium)= 2,2 barn). For samples with mainly incoherent scattering cross sections, the relevant correlation function is the SDF $G_s(r,t)$ which, following Van Hove [25], represents the probability to find the same particle at distance r after a time t.

To overcome the difficulty to collect QENS spectra with a relatively great amount of material [26,27], Doster has proposed an elegant way to get dynamical information by EINS measurements at different resolution values, so taking advantage from the fact that the elastic contribution is at low energy transfer often a factor 100÷1000 higher than the quasi-elastic one [2-4].

In this framework, due to the energy instrumental resolution $\Delta\omega$, the experimentally accessible quantity is the scattering function $S_R(Q,\omega,\Delta\omega)$, *i.e.* the convolution of the scattering law with the instrumental resolution function $R(\omega,\Delta\omega)$ [23,24]. Now, if the resolution in the ω-space is the Dirac delta function, its time Fourier transform, *i.e.* the resolution function in t-space $R(t)$, is a constant with an infinite resolution time, $\tau_{RES}=\infty$, this being the ideal case of elastic neutron scattering.

A resolution function with a non zero but finite characteristic time gives rise to an elastic contribution to which all the motions with a characteristic time τ longer than the resolution time τ_{RES} contribute. On this concern, Doster has shown that the measured EINS intensity profile function can be interpreted as the intermediate scattering function $I(Q,t_R)$, calculated at the instrumental resolution time $t_R=1/\Delta\omega$.

EXPERIMENTAL SECTION

Experimental data were collected by the IN13 spectrometer at ILL which is characterized by a relatively high energy of the incident neutrons (16 meV). The experimental set up was: incident wavelength 2.23 Å; Q-range 0.28÷4.27 Å^{-1}; elastic energy resolution (FWHM) 8 μeV. Raw data were corrected for cell scattering and detector response and normalized to unity at Q=0.28 Å^{-1}. Measurements were performed in the temperature range of 20÷310 K on hydrogenated trehalose and sucrose in H_2O, purchased by Sigma-Aldrich, at a weight fraction value corresponding to 19 water molecules for each disaccharide molecule.

Trehalose and sucrose have the same chemical formula ($C_{12}H_{22}O_{11}$; Mw=342.3), but different structures which could account for the different effectiveness. More precisely sucrose (α-D-glucopyranosil β-D-fructofuranoside) is constituted by a glucose ring (pyranose) in the α configuration and fructose ring (furanose) in the β configuration; the α and β structures of the same monosaccharide differ only in the orientation of the OH groups at same carbon atom in the ring itself (mutarotation equilibria). Trehalose (α-D-glucopyranosil β-D-fructofuranoside) is constituted by two pyranose (six-membered) rings in the same α configuration, linked by a glycosidic bond between the chiral carbon atom C1 of the two rings. Both pure sugars form glasses at temperatures above ambient temperature, but the glass transition temperature T_g of sucrose is significantly lower than that of trehalose ($T_g^{sucrose}$=350K and $T_g^{trehalose}$=388K).

Myoglobin [15] is a small helical protein, closely related to haemoglobin, having the role of intracellular oxygen storage site and consisting of four myoglobin-like subunits that form a tetramer and are responsible for carrying oxygen in blood. Myoglobin data were taken from Ref. [2-4].

RESOLUTION EFFECTS

The scattering law S(Q,ω), linked to the intermediate scattering function I(Q,t) by a time Fourier transform [15,16]:

$$S(Q,\omega) = \frac{1}{\sqrt{2\pi}} \int_{-\infty}^{\infty} I(Q,t)e^{-i\omega t} dt \tag{1}$$

becomes at ω=0:

$$S(Q,\omega = 0) = \frac{1}{\sqrt{2\pi}} \int_{-\infty}^{\infty} I(Q,t) dt \tag{2}$$

Taking into account Eq. (1), the convolution of the scattering law with the instrumental resolution function results:

$$S_R(Q,\omega;\Delta\omega) = \left[\frac{1}{\sqrt{2\pi}} \int_{-\infty}^{\infty} I(Q,t)e^{-i\omega t} dt \right] \otimes R(\omega;\Delta\omega) = \; = \int_{-\infty}^{+\infty} \frac{1}{\sqrt{2\pi}} \int_{-\infty}^{\infty} I(Q,t)e^{-i(\omega-\omega')t} dt R(\omega';\Delta\omega) d\omega' =$$

$$= \int_{-\infty}^{\infty} I(Q,t)e^{-i\omega t} dt \left[\frac{1}{\sqrt{2\pi}} \int_{-\infty}^{\infty} e^{i\omega' t} R(\omega';\Delta\omega) d\omega' \right] = \; = \int_{-\infty}^{\infty} I(Q,t)R(t)e^{-i\omega t} dt \tag{3}$$

Therefore the measured elastic contribution to the scattering is:

$$S_R(Q,\omega = 0;\Delta\omega) = \int_{-\infty}^{\infty} I(Q,t)R(t)\, dt \tag{4}$$

In the ideal elastic case in which the resolution is a delta function in the ω-space, we obtain that the measured scattering function is the scattering law evaluated at ω=0:

$$S_R(Q,\omega = 0;\Delta\omega) = \int_{-\infty}^{\infty} I(Q,t)R(t) dt = \frac{1}{\sqrt{2\pi}} \int_{-\infty}^{\infty} I(Q,t) dt = S(Q,\omega = 0) \tag{5}$$

For the general case evaluation, one can assume for the resolution in the ω-space a Gaussian function, characterized by a resolution time τ_{RES}:

$$R(\omega;\tau_{RES}) = Ce^{-\omega^2 \tau_{RES}^2/2} \tag{6}$$

where $C = a(1 + \tau_{RES})/\sqrt{2\pi}$ in order to guarantee the above discussed limit behaviour for $\tau_{RES} \gg \tau$.

Furthermore, since the main focus here is not to derive specific time dependent properties, let us hypothesize for the intermediate scattering function a Gaussian behaviour with a characteristic relaxation time τ:

$$I(Q,t) = e^{-t^2/2\tau^2} \tag{7}$$

Results that the resolution effect consists in a partial time integration of the intermediate scattering function and:

$$I(Q,t = 0) \le S_R(Q,\omega = 0;\Delta\omega) \le S(Q,\omega = 0) \tag{8}$$

Fig. **1** shows a comparison between $I(Q,t;\tau)$, at a fixed τ value and $R(t;\tau_{RES})$ for different τ_{RES} values. As pointed out by Fig. **1a**, due to the fact that $\tau \ll \tau_{RES}$, it is evident that there is no effect of the resolution function on the measured scattering intensity, *i.e.* $S_R(Q,\omega=0,\Delta\omega)=S(Q,\omega=0)$; by Fig. **1b** it is shown how, in the measured scattering intensity, the resolution function gives rise to a weighted evaluation of the intermediate scattering function; and, finally, in Fig. **1c** it clearly emerges that the resolution function originates the exclusion of a portion of the intermediate scattering function.

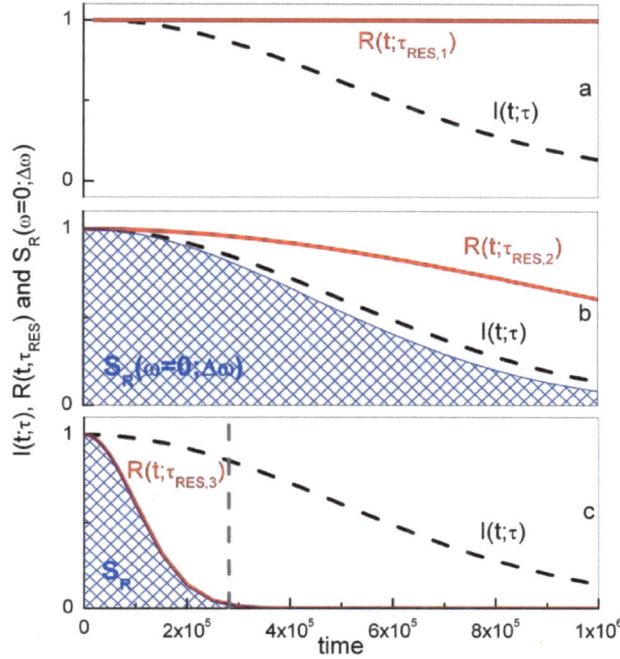

Figure 1: Comparison between $I(t;\tau)$, at a fixed τ value, and $R(t; \tau_{RES})$ for different τ_{RES} values. In Fig. 1a $\tau \ll \tau_{RES}$ and there is no effect of the resolution function on the measured scattering intensity, *i.e.* $S_R(Q,\omega=0,\Delta\omega)=S(Q,\omega=0)$; in Fig. 1b it is shown how, in the measured scattering intensity, the resolution function gives rise to a weighted evaluation of the intermediate scattering function; and, finally, in Fig. 1c it clearly emerges that the resolution function originates the exclusion of a portion of the intermediate scattering function.

Consider Eq. (4) and the theorem of integral average an equivalent time for which the intermediate scattering function and the measured scattering function are proportional, t*, is defined:

$$S_R(Q, \omega = 0; \Delta\omega) \propto I(Q, t^*) \tag{9}$$

Now, evaluating the spatial Fourier transform of the above equation one obtains:

$$F_r\left\{S_R(Q, \omega = 0; \Delta\omega)\right\} \propto G^{self}(r, t^*) \tag{10}$$

The obtained relationship shows that, due to the fact that the SDF can be normalized, the normalized spatial Fourier Transform (F_r) of the measured EINS intensity profile corresponds to SDF evaluated at t*. Therefore, a change in the instrumental energy resolution implies a change of the time at which the SDF is evaluated and, hence, a set of SDFs can be obtained by performing EINS measurements at different energy resolutions. The advantage is based on this relationship that has a general character and does not need any restrictive assumption.

ELASTIC INCOHERENT NEUTRON SCATTERING FUNCTION BEHAVIOUR

The intermediate incoherent neutron scattering function for a system constituted by N particles is well known to be [23,24]:

$$I^{inc}(\underline{Q},t) = \frac{1}{N}\sum_i b_i^{inc\,2} \left\langle \exp\left(i\underline{Q}\cdot\left[\underline{r}_i(t) - \underline{r}_i(0)\right]\right)\right\rangle \tag{11}$$

where the index i runs over the generic i^{th} scattering particle and b_i is the squared scattering length of the ith particle.

Let us operate a partition of the terms contributing to the total intermediate scattering function in groups on the basis of the kind of motion, j, that they can perform:

$$I^{inc}(\underline{Q},t) = \frac{1}{N}\sum_j n_j b_j^{inc\,2} \left\langle \exp\left(i\underline{Q}\cdot\left[\underline{r}_j(t) - \underline{r}_j(0)\right]\right)\right\rangle \tag{12}$$

in which n_j and b^{inc}_j are the number of j type groups and the incoherent scattering length of the jth group.

Let us consider for example a system in which two groups, indexed A and B, which perform vibrational and vibrational plus rotational motions respectively, are present.

In such a case the intermediate scattering function can be rewritten under the form:

$$I^{inc}(\underline{Q},t) = \frac{n_A}{N} b_A^{inc\,2} \left\langle \exp\left(i\underline{Q}\cdot\left[\underline{r}_A(t) - \underline{r}_A(0)\right]\right)\right\rangle = +\frac{n_B}{N} b_B^{inc\,2} \left\langle \exp\left(i\underline{Q}\cdot\left[\underline{r}_B(t) - \underline{r}_B(0)\right]\right)\right\rangle =$$

$$= \frac{n_A}{N} b_A^{inc\,2} \left\langle \exp\left(i\underline{Q}\cdot\Delta\underline{r}_A^V(t)\right)\right\rangle + \frac{n_B}{N} b_B^{inc\,2} \left\langle \exp\left(i\underline{Q}\cdot\left[\Delta\underline{r}_B^V(t) - \Delta\underline{r}_B^R(t)\right]\right)\right\rangle = AI_A^V + BI_B^{V+R} \tag{13}$$

where I_A and I_B represent the intermediate scattering functions of the two groups and where V and V+R refer to the vibration and to the vibration plus rotation contributions, respectively.

When $\Delta\underline{r}_B^V(t) << \Delta\underline{r}_B^R(t)$, i. e. when the vibrational displacement can be considered negligible in respect to the rotational one, it results:

$$I^{inc}(\underline{Q},t) = \frac{n_A}{N} b_A^{inc\,2} \left\langle \exp\left(i\underline{Q}\cdot\Delta\underline{r}_A^V(t)\right)\right\rangle + \frac{n_B}{N} b_B^{inc\,2} \left\langle \exp\left(i\underline{Q}\cdot\Delta\underline{r}_B^R(t)\right)\right\rangle = = AI_A^V + BI_B^R \tag{14}$$

On the other hand, in a more general case, when the two displacements are quantitatively comparable, under the decoupling approximation, we obtain:

$$I^{inc}(\underline{Q},t) = \frac{n_A}{N} b_A^{inc\,2} \left\langle \exp\left(i\underline{Q}\cdot\Delta\underline{r}_A^V(t)\right)\right\rangle + \frac{n_B}{N} b_B^{inc\,2} \left\langle \exp\left(i\underline{Q}\cdot\Delta\underline{r}_B^V(t)\right)\right\rangle\left\langle \exp\left(i\underline{Q}\cdot\Delta\underline{r}_B^R(t)\right)\right\rangle = AI_A^V + BI_B^V I_B^R \tag{15}$$

Now considering the single term contribution and performing a Taylor expansion, we see that an intrinsic deviation from the Gaussian distribution function can be due or to non zero values of the odd expansion terms, which reflect motion distribution asymmetries, or to even terms higher than the second order that are not referable to the second order term.

On the other hand, in the presence of two or more processes, that is of dynamic heterogeneities, under the hypothesis of a Gaussian behaviour for each, one can use a sum of Gaussian contributions, which gives rise to a not Gaussian behaviour, in the analysis of EINS intensity:

$$I^{inc}(Q,t) = \sum_n \frac{n_n b_n^{inc\,2}}{N} e^{-\frac{1}{2}Q^2\left\langle \Delta r^2\right\rangle_n} \tag{16}$$

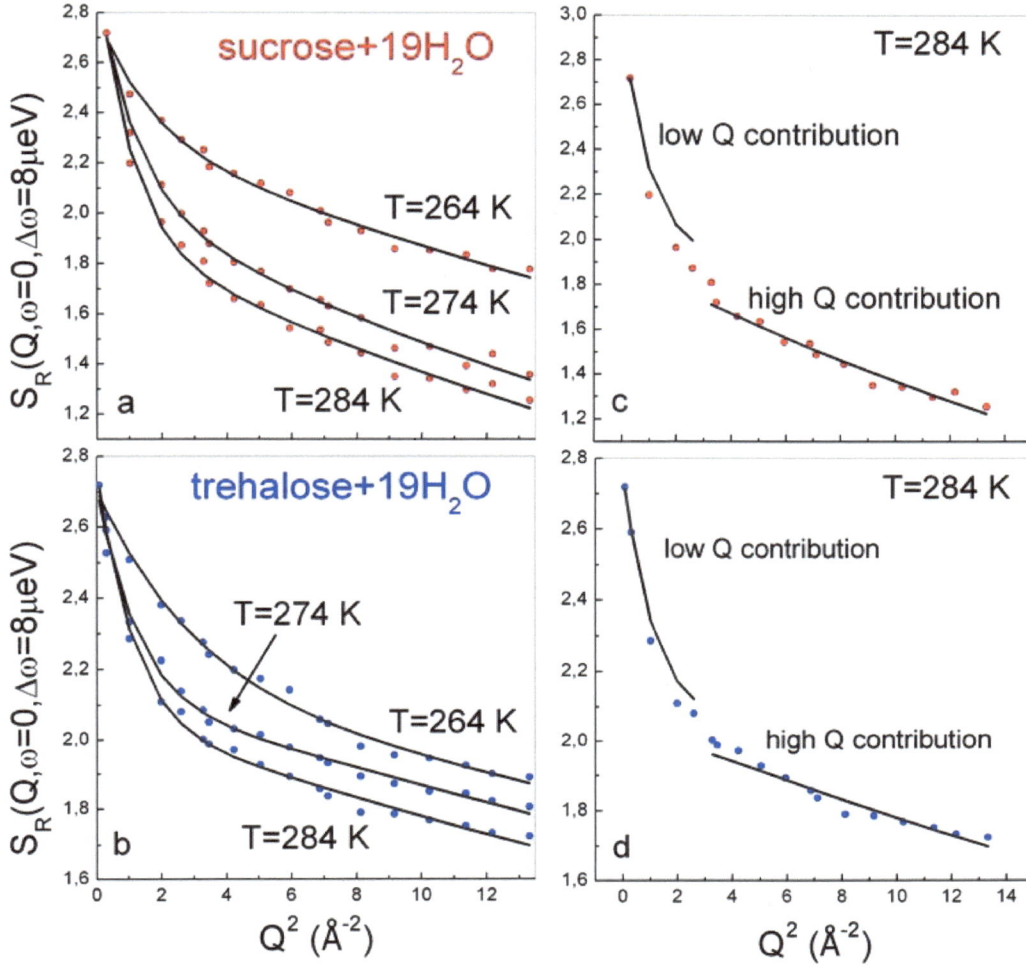

Figure 2: $S_R(Q,\omega=0,\Delta\omega=8\mu eV)$ for sucrose and trehalose, at T=264, 274 and 284 K, together with the fit curves of the two partial contributions (for T=284 K), one relates of low Q domain $(0\div1.7)$ Å^{-1} and the other one of the high Q domain $(1.7\div4)$ Å^{-1}.

The sum of Gaussian contributions in the analysis of measured EINS intensity profile, besides offering advantages from the formal point of view (e.g. the Gaussian is an eigenfunction of the Fourier transform operator), is widely diffused in literature [28] and successfully employed for fitting experimental data taking into account the limits due to their goodness.

In Fig. 2 $S_R(Q,\omega=0,\Delta\omega=8\mu eV)$ and its partial contributions at T=284 K for sucrose and trehalose are shown, respectively. As it can be seen the fitting procedure with Eq. (16) provides two contributions which interest more closely two specific Q ranges, *i.e.* $(0\div1.7)$ Å^{-1} and $(1.7\div4)$ Å^{-1}. One Gaussian function describes the EINS intensity in the high Q domain while the other one in the low Q domain. These two Gaussian functions can be related to different spatial observation windows.

SELF DISTRIBUTION FUNCTION PROCEDURE: DISCUSSIONS AND RESULTS

Following Eq. (10) the normalized space Fourier transform of measured EINS intensity profile corresponds to the SDF evaluated at an equivalent time t*, $G^{self}(r,t^*)$, which is connected to the instrument resolution energy and to the system characteristic time. In the following, the SDF procedure [5-10] is considered taking into account the equivalent time.

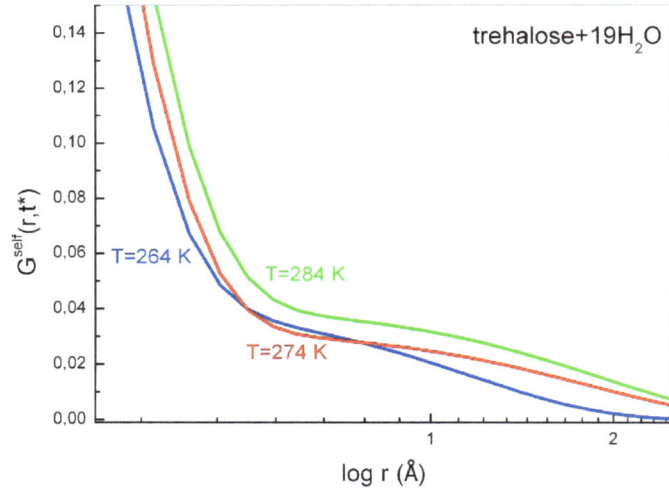

Figure 3: Self-distribution function as a function of r for trehalose at a temperature values of T=264 K, 274 K and 284 K.

In the framework of the SDF procedure, under the single process Gaussian ansatz, we obtain:

$$G^{self}(r,t^*) \propto F_r \left\{ S_R(Q,\omega = 0; \Delta\omega) \right\} = \sum_n A_n F_r \left\{ e^{-Q^2 a_n} \right\} \propto \sum_n A_n G_n^{self}(r,t^*) \tag{17}$$

in which $G_n^{self}(r,t^*)$ are the partial SDFs. To transform the above proportionality in an identity it is sufficient to normalize the total SDF and the single partial SDFs; this implies that:

$$\int_{-\infty}^{\infty} G_n^{self}(r,t^*)dr = 1 \rightarrow G_n^{self}(r,t^*) = 2\left(\pi a_n\right)^{-0.5} e^{-r^2/4a_n} \tag{18}$$

$$A_n \rightarrow B_n = A_n / \sum_n A_n \tag{19}$$

Figs. **3** and **4** show the obtained SDFs as a function of r for trehalose and dry myoglobin in trehalose, respectively, at different temperature values.

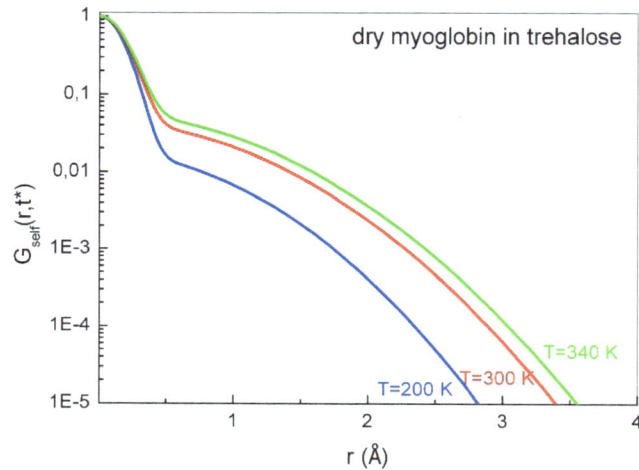

Figure 4: Self-distribution function as a function of r for dry myoglobin in trehalose at a temperature values of T=200 K, 300 K and 340 K.

In this framework the r^2 mean value results:

$$\langle r^2 \rangle = \int_{-\infty}^{\infty} r^2 G^{self}(r,t^*)dr = \sum_n B_n \int_{-\infty}^{\infty} r^2 G_n^{self}(r,t^*)dr = 2\sum_n B_n a_n \qquad (20)$$

In the following we shall consider the comparison between the Gaussian approximation method and the SDF procedure for the MSD determination. As it is well known, following the former the MSD can be obtained by a linear regression in a Guinier plot of the elastic intensity as a function of Q^2 for a set of points that satisfy the inequality $Qx\Delta r<1$, so taking into account the slopes of the measured EINS intensity profile in a Q range starting from $Q=Q_{min}$ up to a given Q' value. Of course at the lowest temperature values, when only a vibrational contribution is present, one can evaluate the intensity profile slope, which is almost constant, in any region of the accessible Q range; this is true also if this contribution decays at $Q>Q_{max}$. In presence of dynamics heterogeneities, starting from a given Q' value and considering smaller and smaller Q' values, the obtained MSD values are dependent on the used Q-range. Now an important point is that the condition $Qx\Delta r<1$ when is applied for evaluating the intensity limit for Q=0 in the presence of dynamic heterogeneities does not allow to use all the accessible Q range since it evaluates locally the limit making use of a restricted region of the measured EINS intensity profile.

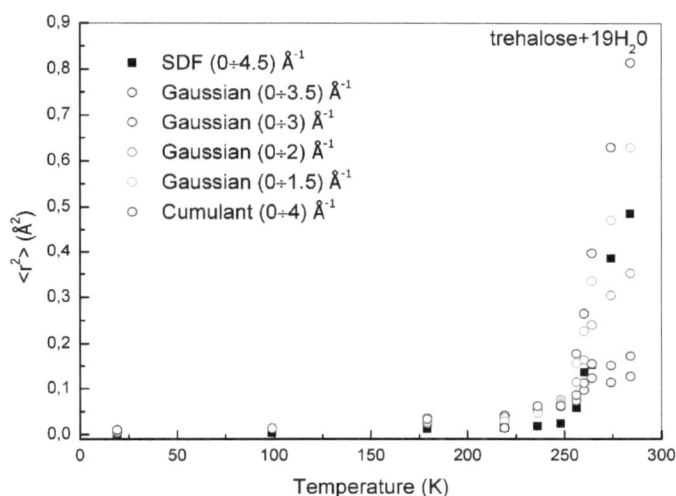

Figure 5: Mean Square Displacement temperature behaviour for trehalose: comparison between Gaussian approximation (different Q range evaluation) and SDF procedure.

The SDF can be applied directly to the experimentally determined EINS profiles as well as to whichever function able to reproduce their behaviour; it represents an integral procedure which takes into account the global Q behaviour and so doing it allows to reduce the error on the Q→0 extrapolation. In addition, the SDF procedure allows to separate the different MSD contributions. Fig. **5** shows the MSD for hydrated trehalose as obtained by applying the SDF procedure and considering the Gaussian approximation for different Q domains. The results obtained with the SDF procedure, in comparison with the Gaussian procedure, are a more harmonic behaviour at low temperature values, the same dynamical transition temperature and a more marked dynamical transition.

Now, let us start to observe that, for a given $Q= Q_{max}-Q_{min}$ range, all the motions decaying to zero for $Q>Q_{min}$ contribute to the elastic incoherent scattering intensity; on the contrary, the motions whose contribution decays in Q within Q_{min} do not contribute.

Figs. **6a** and **6b** show the obtained SDF and its partial contributions as a function of r at T=284K for sucrose and trehalose, respectively. As it can be seen the different kinds of motion are spatially well separated within the accessible Q range; furthermore the SDF very closely follows the first partial SDF in the range (0÷0.5) Å and the second one in the range (0.5÷5) Å. Analogously, in Fig. **7** the obtained SDF and its partial contributions as a function of r for dry myoglobin in trehalose at T=200K are reported.

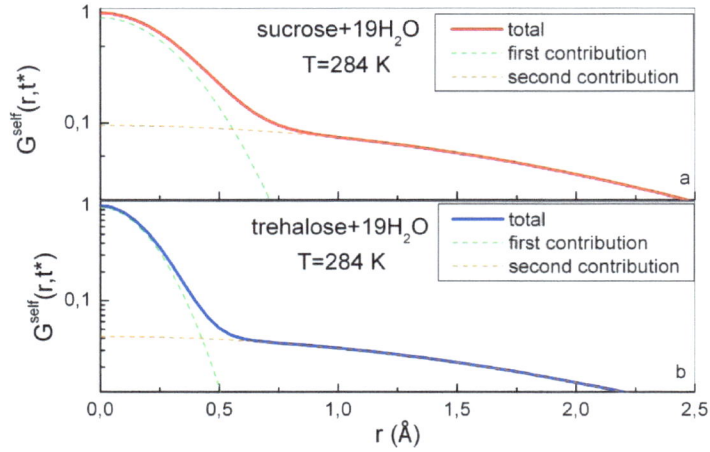

Figure 6: Self-distribution function together their different contributions for sucrose and trehalose at T=284 K.

By performing the spatial Fourier transform of the expression (13) for the intermediate scattering function one obtains:

$$I^{inc}(\underline{Q},t) = AI_A^V + BI_B^{V+R} \rightarrow G^{self}(\underline{r},t) = AG_A^{self,\,V} + BG_B^{self,\,V+R} \tag{21}$$

where the partial SDFs can be associated to the relative scatterer groups. In the case in which the vibrational displacement is negligible respect to the rotational one, the partial SDFs can be related to the different kinds of motion, *i.e.* A=vibration and B=rotation:

$$I^{inc}(\underline{Q},t) = AI_A^V + BI_B^R \rightarrow G^{self}(\underline{r},t) = AG_A^{self,\,V} + BG_B^{self,\,R} \tag{22}$$

On the other hand, in the general case in which the above condition cannot be applied, under a decupling hypothesis for rotational and vibrational motions one obtains:

$$I^{inc}(\underline{Q},t) = AI_A^V + BI_B^V I_B^R \rightarrow G^{self}(\underline{r},t) = AG_A^{self,\,V} + BG_B^{self,\,V} \otimes G_B^{self,\,R} \tag{23}$$

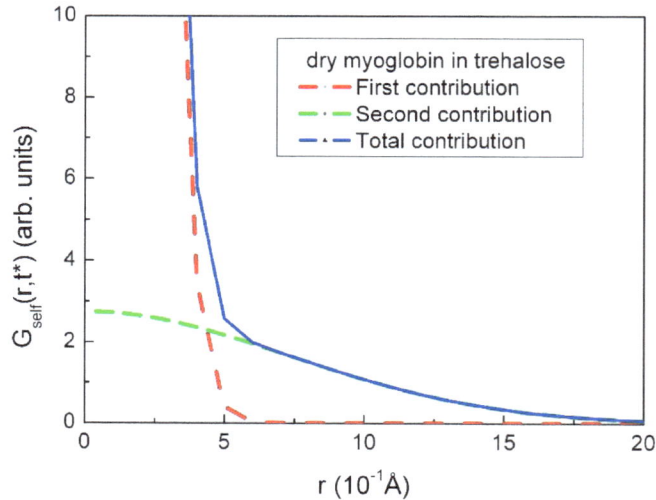

Figure 7: Self-distribution function together their different contributions for myoglobin in trehalose at T=200 K.

It is possible now to obtain the partial MSD values:

$$\left\langle r^2 \right\rangle_n = \int_{-\infty}^{\infty} r^2 G_n^{self}(r,t^*)dr = 2a_n \tag{24}$$

the exponent of each Gaussian being the MSD relative to a particular r-domain and the weight A_n being interpretable as the relative percentage weight. Therefore this procedure allows to obtain the autocorrelation function $G_{self}(r,t^*)$ versus r, together with its different partial contributions, as well to determine the partial MSDs, their weights and the total MSD. Figs. **8a** and **8b** shows the partial MSDs related for sucrose and trehalose, evaluated by the SDF procedure, in the temperature range of 20÷287K to the high-r domain and to small-r domain, respectively. As it can be seen, the partial MSD behaviours of sucrose and trehalose are equivalent in the high r-domain, whereas they are different in the small-r domain.

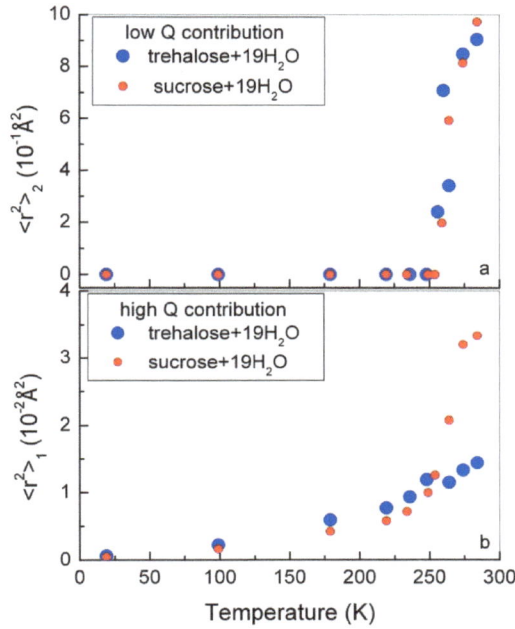

Figure 8: Mean Square Displacement temperature behaviour for trehalose and sucrose spatial observation window analysis.

This circumstance suggests that the higher structure sensitivity of sucrose in respect to trehalose should be related to the small spatial observation windows. It is also important to observe that the dynamical transition temperature of the partial MSDs is equal each other and is equal to the dynamical transition temperature of the average MSD. Eq. (24) can be also expressed by:

$$\left\langle r^2 \right\rangle = 2\sum_n B_n a_n = \sum_n B_n \left\langle r^2 \right\rangle_n \tag{25}$$

As it can be seen, the MSD is not the simple sum of the different displacement contributions but corresponds to a weighed sum of the MSD contributions associated with the different relaxations in which the weights are obtained by the fitting procedure of measured EINS intensity data.

We can evaluate the MSD by using an alternative way:

$$\left\langle r^2 \right\rangle = -2 \left. \frac{\partial I(Q,t)}{\partial Q^2} \right|_{Q=0} = \tag{26}$$

$$= -2\frac{\partial \sum_n B_n e^{-\frac{1}{2}Q^2\langle \Delta r^2\rangle_n}}{\partial Q^2}\Bigg|_{Q=0} = -2\frac{\partial\left(B_1 e^{-\frac{1}{2}Q^2\langle \Delta r^2\rangle_1} + ... + B_n e^{-\frac{1}{2}Q^2\langle \Delta r^2\rangle_n}\right)}{\partial Q^2}\Bigg|_{Q=0} =$$

$$= \frac{\langle \Delta r^2\rangle_1 A_1 e^{-\frac{1}{2}Q^2\langle \Delta r^2\rangle_1} + ... + \langle \Delta r^2\rangle_n A_n e^{-\frac{1}{2}Q^2\langle \Delta r^2\rangle_n}}{A_1 + ... + A_n}\Bigg|_{Q=0} = \tag{27}$$

$$= \frac{\langle \Delta r^2\rangle_1 A_1 + ... + \langle \Delta r^2\rangle_n A_n}{A_1 + ... + A_n} = \langle \Delta r^2\rangle_1 B_1 + ... + \langle \Delta r^2\rangle_n B_n = \sum_n B_n \langle \Delta r^2\rangle_n \tag{28}$$

Therefore the MSD obtained by SDF procedure represents a good extrapolation to $Q=0$ of the EINS intensity slope. As it can be seen in Fig. **9**, by decreasing the Q^2 range the slopes pass from an almost constant value towards a monotonic increase which provide, by their extrapolation to zero, the half of the MSD.

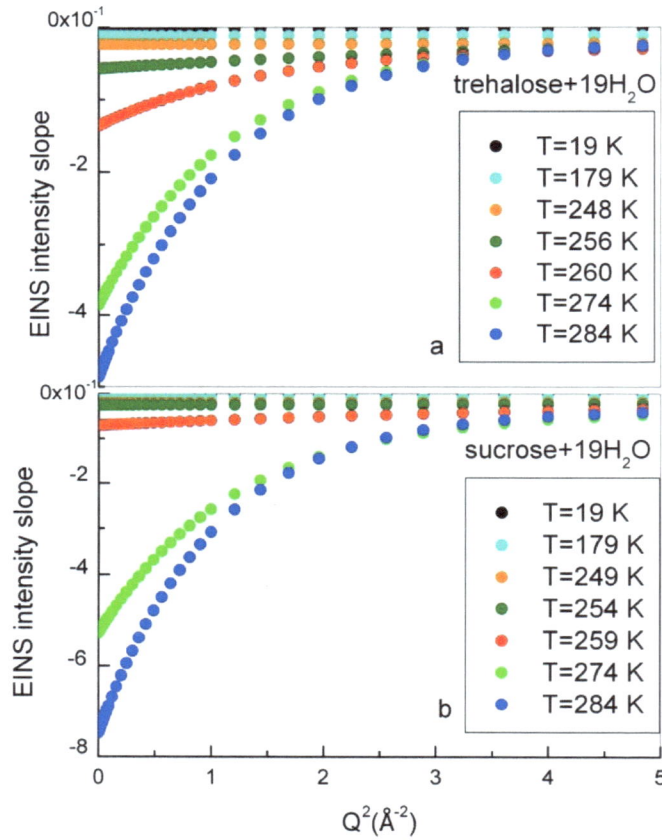

Figure 9: Elastic Incoherent Neutron Scattering intensity slopes as a function of Q^2 for trehalose and sucrose.

CONCLUSIONS

In the present chapter the evaluation of the resolution effect on the measured EINS intensity profile is firstly considered. It is shown that the resolution effect consists in the partial time integration of the intermediate scattering function. We considered the limit cases when $\tau << \tau_{RES}$ and obtains that $S_R(Q,\omega=0,\Delta\omega)=S(Q,\omega=0)$. We have also defined an equivalent time t*, for which the normalized spatial Fourier Transform of the measured EINS intensity profile corresponds to the SDF evaluated at t=t*. On this concern, starting from the general form of the intermediate incoherent scattering function it is possible to introduce a fitting expression for measured EINS intensity profile containing the average displacements associated with the different spatial domains together with their relative weights. A comparison between the SDF and the Gaussian procedure is presented.

By applying the SDF procedure to water/homologous disaccharide mixtures, a different dynamical behaviour in the large Q range region is pointed out at temperature values higher than about 250 K. It emerges that the hydrogen bond imposed network of the water-trehalose mixture appears to be stronger in respect to that of the water-sucrose mixture and this result can justify the highest bioprotectant effectiveness of trehalose in comparison with sucrose.

REFERENCES

[1]   Alefeld B, Kollmar A, Dasannachrya BA. The one-dimensional CH$_3$-quantumrotator in solid 4-methyl-pyridine studied by inelastic neutron scattering. J Chem Phys 1976; 63: 4415-7.
[2]   Doster W, Diehl M, Gebhardt R, Lechner RE, Pieper J. TOF-elastic resolution spectroscopy: time domain analysis of weakly scattering (biological) samples. Chem Phys 2003; 292: 487-94.
[3]   Diehl M, Doster W, Petry W, Schober H. Water-coupled low-frequency modes of myoglobin and lysozyme observed by inelastic neutron scattering. Biophys J 1997; 73: 2726-32.
[4]   Doster W, Diehl M, Petry W, Ferrand M. Elastic resolution spectroscopy: a method to study molecular motions in small biological samples. Physica B 2001; 301: 65-8.
[5]   Magazù S, Maisano G, Migliardo F, Benedetto A. Mean square displacement evaluation by elastic neutron scattering self-distribution function. Phys Rev E 2008; 77: 0618021-6.
[6]   Magazù S, Maisano G, Migliardo F, Benedetto A. Mean square displacement from self-distribution function evaluation by elastic incoherent neutron scattering. J Mol Struct 2008; 882: 140-5.
[7]   Magazù S, Maisano G, Migliardo F, Benedetto A. Elastic Incoherent Neutron Scattering on systems of biophysical interest: mean square displacement evaluation from self-distribution function. J Phys Chem B 2008; 112: 8936-42.
[8]   Magazù S, Maisano G, Migliardo F, Galli G, Benedetto A, Morineau D, Affouard F, Descamps M. Characterization of molecular motions in biomolecular systems by elastic incoherent neutron scattering. J Chem Phys 2008; 129: 1551031-8.
[9]   Magazù S, Maisano G, Migliardo F, Benedetto A. Biomolecular motion characterization by a self-distribution-function procedure in elastic incoherent neutron scattering. Phys Rev E 2009; 79: 0419151-9.
[10]   Magazù S, Maisano G, Migliardo F, Benedetto A. Motion characterization by self- distribution-function procedure. Biochim Biophys Acta 2010; 1804: 49-55.
[11]   Magazù S. IQENS - dynamic light scattering complementarity on hydrogenous systems. Physica B 1996; 226: 92-106.
[12]   Affouard F, Bordat P, Descamps M, Lerbret A, Magazù S, Migliardo F, Ramirez-Cuesta AJ, Telling MFT. A combined neutron scattering and simulation study on bioprotectant systems. Chem Phys 2005; 317: 258-66.
[13]   Magazù S, Migliardo F, Telling MFT. α,α-Trehalose−Water Solutions. VIII. Study of the Diffusive Dynamics of Water by High-Resolution Quasi Elastic Neutron Scattering. J Phys Chem B 2006; 110: 1020-5.
[14]   Magazù S, Villari V, Migliardo P, Maisano G, Telling MTF. Diffusive Dynamics of Water in the Presence of Homologous Disaccharides: A Comparative Study by Quasi Elastic Neutron Scattering. IV. J Phys Chem B 2001; 105: 1851-5.
[15]   Kendrew JC, Dickerson RE, Strandberg BE, Hart RG, Davies DR, Phillips DC, Shore VC. Nature 1960; 185: 422-6.
[16]   Cusack S, Doster W. Temperature dependence of the low frequency dynamics of myoglobin. Measurement of the vibrational frequency distribution by inelastic neutron scattering. Biophys J 1990; 58: 243-51.
[17]   Loncharich RJ, Brooks BR. Temperature dependence of dynamics of hydrated myoglobin: Comparison of force field calculations with neutron scattering data. J Mol Biol 1990; 215: 439-55.
[18]   Smith J, Kuczera K, Karplus M. Dynamics of myoglobin: comparison of simulation results with neutron scattering spectra. Proc Natl Acad Sci USA 1990; 87: 1601-5.
[19]   Ahn JS, Kanematsu Y, Enomoto M, Kushida T. Determination of weighted density of states of vibrational modes in Zn-substituted myoglobin. Chem Phys Lett 1993; 215: 336-40.
[20]   Henry ER, Eaton WR, Hochstrasser RM. Molecular dynamics simulations of cooling in laser-excited heme proteins. Proc Natl Acad Sci USA 1986; 83: 8982-6.
[21]   Cordone L, Ferrand M, Vitrano E, Zaccai G. Dehydration and crystallization of trehalose and sucrose glasses containing carbonmonoxy-myoglobin. Biophys J 1999; 76: 1043-7.

[22] Cottone G, Cordone L, Cicciotti G. Molecular Dynamics simulation of carboxy-myoglobin embedded in a trehalose-water matrix. Biophys J 2001; 80: 931-8.

[23] Bee M. Quasielastic Neutron Scattering. Adam Hilger, Bristol, 1988.

[24] Volino F. Spectroscopic methods for the study of local dynamics in polyatomic fluids, in: Dupuy J, Dianoux AJ (Eds.). Microscopic structure and dynamics of liquids, New York, Plenum Press, 1978.

[25] Van Hove L. Correlations in space and time and Born approximation scattering in systems of interacting particles. Phys Rev 1954; 95: 249-62.

[26] Doster W, Cusack S, Petry W. Dynamical transition of myoglobin revealed by inelastic neutron scattering. Nature 1989; 337: 754–6.

[27] Doster W, Settles M. Protein–water displacement distributions. Biochim Biophys Acta 2005; 1749: 173-86.

[28] Becker T, Smith JC. Energy resolution and dynamical heterogeneity effects on elastic incoherent neutron scattering from molecular systems. Phys Rev E 2003; 67: 0219041-8.

SECTION II

Dynamics of Biological Molecules

CHAPTER 3

Dynamics of Model Membranes

Francesca Natali[1,*] and Marcus Trapp[2]

[1]CNR-IOM, OGG, c/o Institut Laue-Langevin, Grenoble, FR-38000, France and [2]Institut de Biologie Structurale J.-P. Ebel, UMR 5075, CNRS-CEA-UJF, Grenoble, FR-38027, France

Abstract: Biological membranes are complex multicomponent systems whose dynamics and structure provide their physiological function. Many parameters interplay to determine the membrane flexibility; among them, lipid composition, lipid-protein interaction, hydration, temperature etc.

We provide here a tentative overview of recent successful neutron scattering experiments on different oriented model membranes, with the aim to demonstrate the many unique advantages that elastic and quasi-elastic neutron scattering offer for the investigation of membrane dynamics.

INTRODUCTION

Membranes are fundamental elements of all living cells, performing important active functions controlling transport of molecules and ions across them. The lipid bilayer itself is a complicated multicomponent system made up of about 100 different lipids, differing in the hydrocarbon chain and in the polar head groups. Structural properties of these bilayers are very important for their functional activity and have been investigated in detail [1]. Much less is known on bilayer dynamics.

On the other hand, it is well acknowledged that the different kinds of motions of the lipid chains are of paramount importance for the physiological function of membranes and those motions can be modulated by a number of factors like lipidic composition and protein interactions.

It is commonly accepted that at least six different lipid movements can be observed in oriented membranes, characterized by specific time scale windows [2]. In the 10^{-11}-10^{-8} s range the rotational diffusion about the lipid molecular axis and the chain defect motions can be observed. In the latter, the lipid chain oscillates forming an angle θ with respect to the molecule axis, along the normal to the membrane plane (wobbling motion). The origin of this kind of motion is still not completely understood and could be due to a fluctuating cis-trans isomerization of the acyl chains. On the other hand, the 10^{-10} – 10^{-9} s time scale is characterized by the exhibition of the vertical (*i.e.* parallel to the membrane normal) vibrational motion of the lipid molecules. At lower times, lateral diffusion in the bilayer plane (10^{-9} s) and rotational and flip flop motion of the lipid headgroups (3×10^{-9} s), take place. Finally, much slower membrane dynamics is promoted by the collective modulations of the bilayers, which, observed in the s time scale, conferees the membrane roughness.

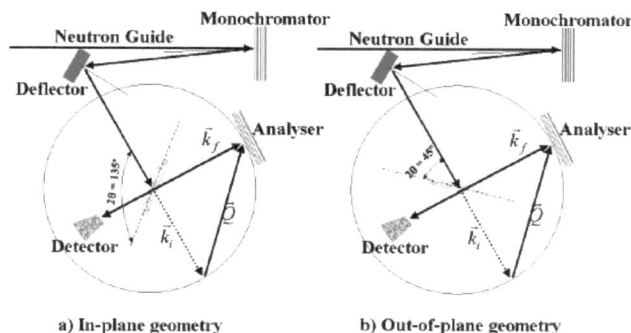

a) In-plane geometry   b) Out-of-plane geometry

Figure 1: By rotating the sample with respect to the incoming beam, the momentum transfer is predominantly oriented parallel (135°, panel a) and perpendicular (45°, panel b) to the membrane surface.

*Address correspondence to Francesca Natali: CNR-INFM, OGG, c/o Institut Laue-Langevin, Grenoble, FR-38000, France; E-mail: natali@ill.fr

Salvatore Magazù and Federica Migliardo (Eds)

Incoherent neutron scattering is a powerful technique to investigate dynamics of biological molecules, which are characterized by an extremely high hydrogen content, representing at least half of the total number of atoms of the system. In fact, the hydrogen incoherent neutron scattering cross section is about an order of magnitude larger than that of other atomic species typically present in biological systems. Thus, incoherent neutron scattering provides averaged information on the global dynamics of the system. In particular, elastic (ENS) and quasi-elastic incoherent neutron scattering (QENS) represent an optimal tool to provide unique information on lipid dynamics [3-5]. For these measurements, highly ordered samples are required in order to allow selection of the direction of momentum transfer with respect to the membrane normal, and therefore separation of in-plane and out-of-plane motions (Fig. 1).

This is achieved using oriented lipid multilayer obtained drying a thin layer of liposome suspension on a solid support, typically SiO_2 polished wafers.

We report here few selected examples of the investigation of dynamical properties of different membrane model systems.

EFFECT OF MYELIN BASIC PROTEIN ON THE DYNAMICS OF ORIENTED LIPID BILAYERS

Myelin is the discontinuous multibilayer membrane sheath wrapped around the nerve axon. The integrity of the myelin sheath is fundamental to optimize the action potential conduction along the axon [6,7]. It contains proteins which are believed to play an important role in maintaining the membrane stack order [8-11]. Among them, the Myelin Basic Protein (MBP) is the second most abundant myelin protein that counts for up to 30% of the total protein fraction in the Central Nervous System (CNS) [12].

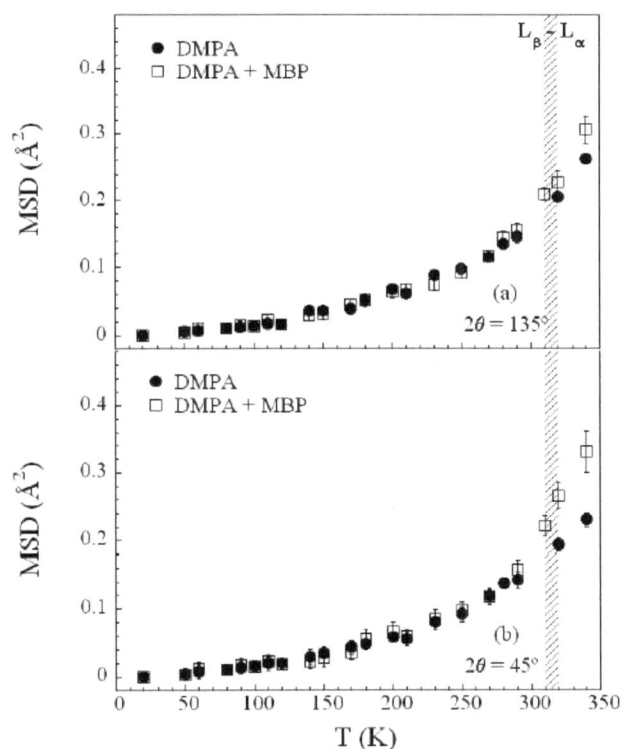

Figure 2: Temperature dependence of the normalized Mean Square Displacements (MSD) of DMPA (filled symbols) and DMPA+MBP (empty symbols). Incoherent elastic neutron scattering data have been acquired at the high resolution backscattering spectrometer IN13 at the Institut Laue-Langevin in Grenoble - France (ILL), using an energy resolution of 8µeV, corresponding to ~100 ps time scale accessible. MSD are calculated, in agreement to the Gaussian model, as the slope of the logarithm of the elastic intensity versus squared momentum transfer Q^2, in the low-Q range fulfilling the condition of the Guinier approximation. (a) and (b) refer to the in plane (135°) and out-of-plane (45°) configurations, respectively. The lipid phase transition from gel (L_β) to liquid-crystalline (L_α), occurring at T_c close to 320 K, is also shown [16].

MBP is an extrinsic protein that, when removed from its native environment in the membrane and isolated in the water-soluble form, appears as an extended, flexible, irregular coil having little secondary structure [8,13]. However, several studies have demonstrated that MBP, according to its structural role, can interact with specific lipids to form ordered assemblies [12,14,15]. On the other hand, it is still not clear whether the absence of MBP affects the lipid dynamics. Recent elastic and quasi-elastic neutron scattering experiments, performed on highly oriented dimyristoyl phosphatidic acid (DMPA) phospholipid multilayers, simulating the myelin sheath, have demonstrated that it is possible to separate the in-plane and out-of-plane contributions of the membrane dynamics, thus to investigate the eventual appearance of membrane anisotropy [16]. It was also observed that the addition of the MBP to the DMPA membranes affects significantly the membrane dynamics. In particular the lipid mobility in the out-of-plane configuration (*i.e.* at 45° with respect to the incident beam) was shown to increase (Fig. **2**).

Moreover, the lipids are known to exhibit a complex temperature-dependent phase diagram, with at least two phase transitions from crystalline (L_c-phase), to gel (L_β-phase), and then to liquid-crystalline (L_α-phase) structures, reflecting different degrees of disorder. The transition from gel to liquid-crystalline phase ($L_\beta \rightarrow L_\alpha$) in saturated lipids normally occurs at temperatures around $T_c \sim 320$ K, depending on the membrane composition. Thus, the observed enhanced lipid flexibility in DMPA+MBP at $T > 310$ K (Fig. **2**), that leads to higher MSD values, could be assigned to the effect of the membrane structural modification occurring across the lipid phase transition. Moreover, the out-of-plane membrane dynamics is markedly characterized by spatially restricted vertical diffusive motions of the lipids, significantly enhanced by the MBP above the gel to liquid-crystalline ($L_\beta \rightarrow L_\alpha$) DMPA phase transition (Fig. **3**). On the other hand, the in-plane dynamics seems to involve predominantly the spatially restricted lateral diffusion of the lipids on the membrane surface that appears to be only slightly affected by the presence of the MBP.

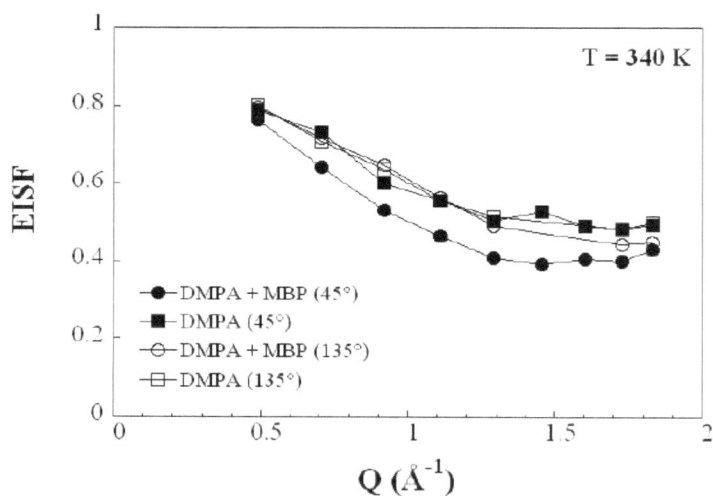

Figure 3: Q dependence of the Elastic Incoherent Structure Factor (EISF) of DMPA and DMPA+MBP, measured in the L_α liquid phase (T = 340 K) and at both geometrical configurations (135° and 45°). Incoherent quasi-elastic neutron data have been acquired at the IN16 backscattering spectrometer at ILL, using an energy resolution of 0.9μeV, corresponding to ~2 ns time scale accessible. The presence of an EISF indicates that the observed motions have a localized diffusive nature. The EISF values at high-Q show a tendency to an asymptotic non-zero value (close to 0.5), suggeststsing that about 50% of the hydrogen contributing to the spectrum perform only fast vibrational motions and do not contribute to the quasi-elastic part of the spectrum.

DYNAMICS OF LIPOPLEX-DNA GENE VECTORS

Human gene therapy is defined as the transfer of nucleic acids to somatic cells of a patient providing a therapeutic effect [17,18]. Among the non-viral carriers, cationic liposomes (CL) attract a significant interest because of their unique properties and their efficiency in acting as vehicles for DNA delivery into eukariotic cells [18,19].

Many theoretical and experimental studies have been performed [20-24] to understand the factors governing the energetic, structural, and thermodynamic characteristics of CL-DNA complexes; these properties, strongly influenced by the specific composition of lipoplexes, are essential for optimizing their transfection efficiency. The cationic lipoplexes are normally constituted by the mixture of a neutral (helper) lipid and a charged lipid; the former determines a given structure (in particular lamellar or hexagonal geometry), whereas the latter is fundamental for delivering the genes into the cell [22,23-30].

The direct interaction of the positively charged lipid headgroup with the negatively charged phosphate of the DNA backbone is suggested by many authors to be the main mechanism for the complexation of DNA with cationic lipids [26,28]. This is also supported by both fluorescence techniques and differential scanning calorimetry [25] suggesting that the release of bound water and counterions is the driving force behind complex formation [26]. Besides, Choosakoonkriang and coworkers suggest that complexation with DNA induces a small increase in the disordered conformation of the lipid alkyl chain, altering the packing of the lipid (due to the alignment of the lipid headgroup with the DNA phosphate) and determining a greater fluidity of the apolar region of the membrane [27,28].

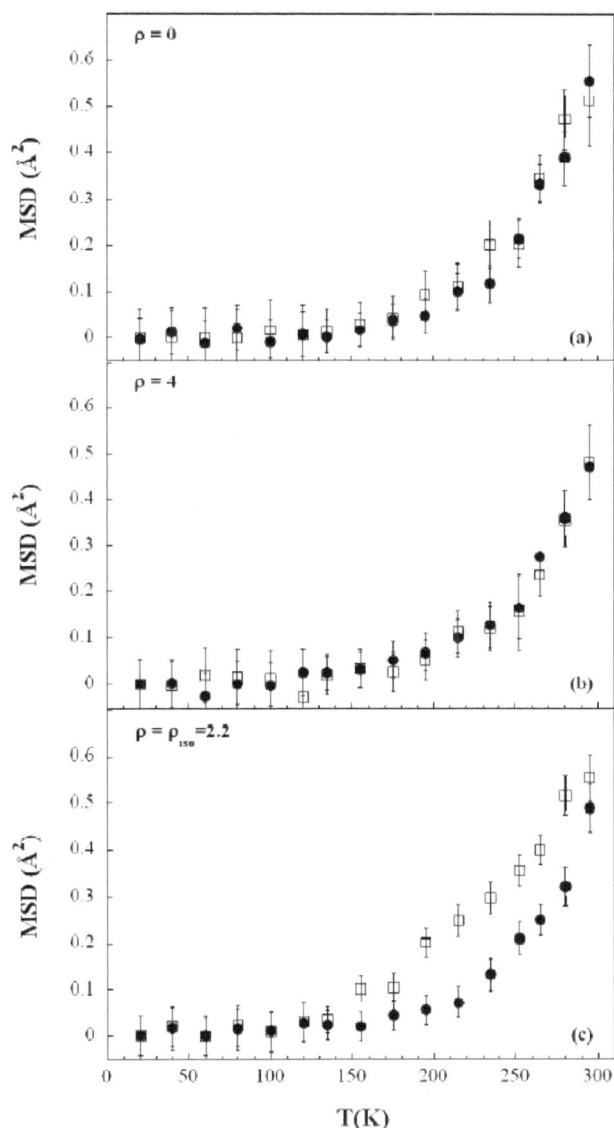

Figure 4: Normalized MSD of the CLs-DNA mixed multilayers *vs.* T, as a function of the cationic lipid/DNA molar weight ratio ρ. Panel a: pure lipids (ρ=0); panel b: excess of liposomes (ρ=4); panel c: isoelectric point (ρ=ρ$_{iso}$=2.2). Open squares: out-of-plane direction; filled circles: in-plane direction. Incoherent elastic neutron scattering data have been acquired at IN13-ILL, with an energy resolution of 8μeV [31].

Unfortunately, very few studies have been performed on the dynamics of CLs-DNA systems to date; among them, is the recent investigation of highly oriented lamellar CLs-DNA complexes consisting of calf thymus DNA added to 1:1 ratio binary mixtures of cationic monovalent lipid DOTAP (dioleoyl trimethylammonium propane) and the neutral helper lipid DOPC (dioleoyl phosphatidylcholine), both lipids having two 18-carbon (C18) aliphatic chains per molecule [31].

The membrane dynamics is shown to be strongly dependent on the cationic lipid/DNA molar weight ratio ρ (Fig. **4**). Indeed, CLs-DNA at the isoelectric point (Fig. **4**, panel c), displays a marked anisotropy in the mean square displacements. In particular, higher dynamics is observed in the out-of-plane configuration.

The main result is that a minimum amount of DNA phosphate groups is not sufficient to induce modifications in membrane dynamics. On the other hand, at the isoelectric point, the balance of the total net charge inside the complex, together with the displacement of bound water molecules into solution (accompanied by counterion release), provides new degrees of freedom to the lipoplex. This enhances the apolar region fluidity and results in a very large increase of the out-of-plane lipid motions, mainly assigned to spatially confined vertical translation of the entire lipid molecule.

An exhaustive description of the results, together with a proposed model to interpret the data, is reported elsewhere [31].

More recently [32], the investigation was extended to DOTAP-DOPC model membranes as a function of the DOTAP/(DOPC+DOTAP) ratio ϕ (Fig. **5**). The study revealed a reduction of the dynamics along the direction normal to the membrane induced by increasing the neutral lipid (DOPC) concentration from $\phi = 0.5$ to 0.8.

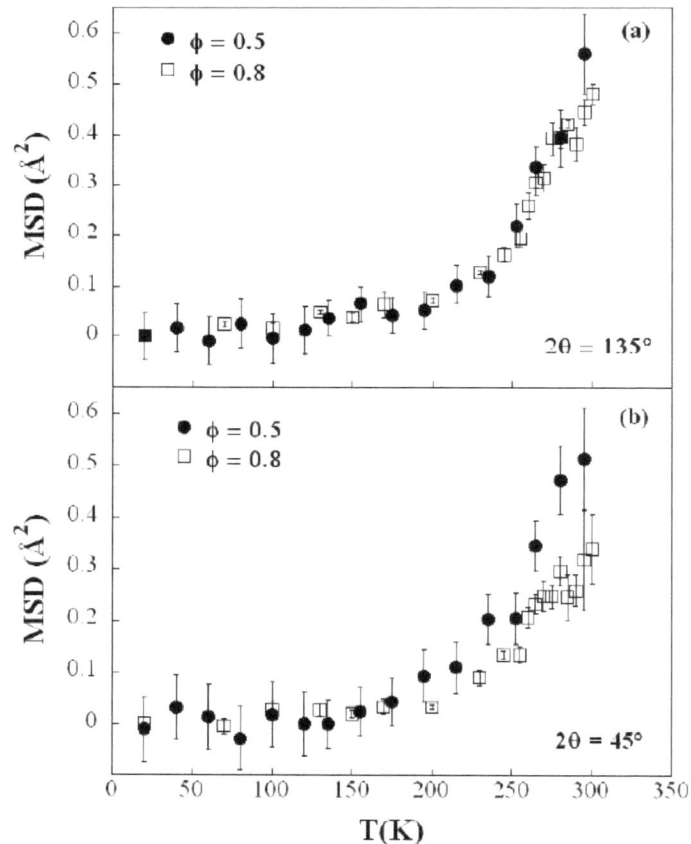

Figure 5: Temperature dependences of MSD for $\phi = 0.5$ (filled circles) and $\phi = 0.8$ (empty squares), measured at 135° (a) and 45° (b) with respect to the membrane normal. Incoherent elastic neutron scattering data have been acquired at IN13-ILL, with an energy resolution of 8μeV [32].

The explanation of the observed behaviour has to take into account the origin of the detected elastic intensity. While in the case of $\phi = 0.5$ equal contributions arise from DOPC and DOTAP scattering, the $\phi = 0.8$ sample reflects a difference of the relative weight in the total revealed signal. Indeed, the 0.8 lipid ratio refers to the increasing of DOPC concentration of a factor of 4 with respect to the DOTAP. Thus, being the density cross section comparable for the lipids, in the $\phi = 0.8$ case the 80 % contribution to the total signal is assigned to the DOPC, while only 20 % arises from DOTAP.

On the other hand, DOPC has greater hydrophilic heads, characterized by the presence of a phosphate PO_4^- group not present in the cationic DOTAP where there is only the aminic one NH_3^+. Thus, the decrease of the out-of-plane mean square displacements correlated to the increasing DOPC concentration may be interpreted in terms of more localized vertical diffusion due to less free space accessible (Fig. **6**).

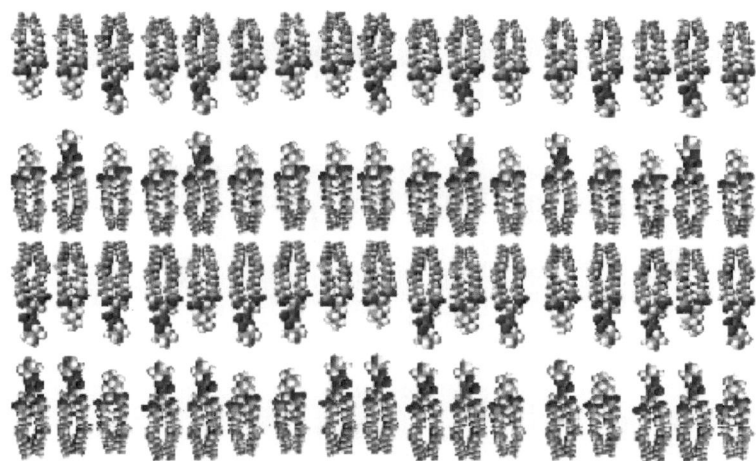

Figure 6: Schematic representation of the DOTAP+DOPC oriented lipid mixture. The greater occupation of the interbilayer free space assigned to the DOPC, is evident.

Moreover, the ϕ independent in-plane dynamics confirms that the key-role in the changes of membrane dynamics is mainly governed by the different size of the lipid heads.

THE DYNAMICS OF THE GANGLIOSIDES IN BILAYER DOMAINS

Many macromolecules of biological relevance are characterized by the presence of sugar moieties with different degrees of complexity. A particular class of sugar-containing molecules is that of gangliosides, glycosphingolipids abundant in neuronal plasma membranes, which are believed to play a role in a number of cellular functions, including cell recognition, adhesion, regulation, signal transduction, and development of tissues. They are predominantly located in the outer leaflet of the membrane and may act to protect the membrane from harsh conditions such as low pH or degradative enzymes [33, 34]. Ganglioside are amphiphilic molecules constituted of a ceramide and a saccharidic headgroup including one or more charged sugars (sialic acid). The mechanical properties and biological functions of gangliosides are strongly dependent on the behaviour of the lipids to which they are bound [35-38]. The properties of lipids are then likely to be strongly affected by the microdomain presence and arrangement.

One of the most commonly studied gangliosides is galactosyl-Nacetylgalactosaminyl(N-acetyl-neuraminyl)galactosylglucosylceramide (GM1) [39]. GM1 is a member of the glycosphingolipids family and contains four neutral sugar residues and a negatively charged sialic acid residue. The glycolipid monosialoganglioside GM1 is widely distributed in all tissues and reaches its highest concentrations in the central nervous system. It is primarily located in the outer surface of the mammalian cell's plasma membrane and in synaptic membranes of the CNS. GM1 ganglioside modulates a number of cell surface and receptor activities as well as neuronal differentiation and development, protein phosphorilation and synaptic function.

Recently the formation of ganglioside GM1-rich domains in monolayers and bilayers is an area of increased scientific interest. In particular, specialized membrane domains composed of phospholipids, glycolipids, and cholesterol – so called lipid rafts - are thought to play a role in a diverse range of processes. Among them membrane trafficking and signaling through specific membrane protein interactions, where the raft microdomain acts as a platform for various cellular events [40-48].

The first investigation of the dynamics of low-hydration lamellar systems containing gangliosides concerned the effect induced by the presence of minority amount of GM1 molecules on the dynamics of oriented lamellar DMPC assemblies deposited on Si flat substrate [49].

A strong gap in the incoherent elastic neutron intensity, measured on IN13 at ILL, was observed across the gel-to-liquid lipid phase transition region ($T_c \sim 320K$, at the given membrane hydration) (Fig. 7).

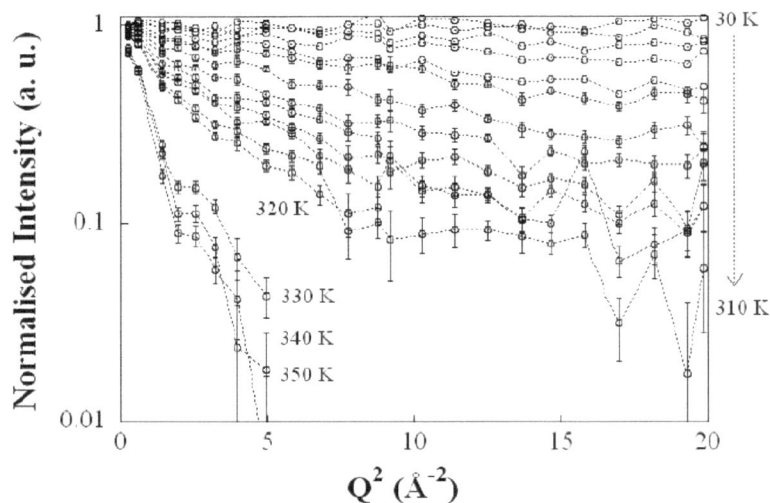

Figure 7: Normalized elastic intensity of in-plane DMPC *vs.* Q^2, measured on IN13-ILL. Temperatures from 30 K to 350 K are reported. High temperature data points at high Q values are omitted for clarity, due to the large error bars associated to the low signals recorded [49].

Below the lipid phase transition, the in-plane lipid dynamics is sensibly increased upon addition of GM1, while a strong reduction of the out of plane mobility is observed across the phase transtion (data not shown, publication in progress, [49]).

At a glance, the presence of GM1 results in a clear increase in anisotropy, damping the mean square displacement in out-of-plane direction at $T > T_c$, and in a kink mainly affecting the in-plane curve at temperatures slightly lower than 300 K, as if a double process is taking place. The presence of domains enriched in gangliosides could provide the clue for a reasonable interpretation of these results. Moreover, the present results would support the hypothesis, drawn on the basis of past experiments with various techniques evidencing their ability to establish an extended network of interactions, that gangliosides play a central role in the coordination of the structure and dynamics of their environment.

INFLUENCE OF HYDRATION ON THE DYNAMICS OF MODEL MEMBRANE SYSTEMS

Biological membranes are composed not only of different kinds of lipids but also of membrane proteins and molecules like e.g. cholesterol and ethanol. Lipid membranes consisting of only one type of lipid such as 1,2-Dimyristoyl-*sn*-Glycero-3-Phosphocholine (DMPC) serve as role models for their more complex counterparts in biological systems. The phase behaviour is strongly dependent on the hydration level of the membranes [50]. Inelastic neutron scattering (INS) [51], quasi elastic neutron scattering (QENS) [2,52,53] and neutron spin echo spectroscopy (NSE) [54] have already been employed to study local as well as collective dynamics of these membranes. However, most of these studies lack a systematic investigation of the behaviour of the model membranes in dependence on their hydration.

By hydrating the mulilayer stack using pure D_2O or saturated salt solutions, the relative humidity (rh) can be adjusted and thus the level of hydration of the membranes.

From structural investigations [50] it is known that a lower degree of hydration causes a shift of the main phase transition to higher temperatures. As a result of the lower water content the repeating distance of the bilayers is reduced. The effect can be seen e.g. by neutron diffraction as a shift in the distance of the Bragg peaks originating from the lipid bilayers as shown in Fig. **8**. *Via* the relation $d=\lambda/2*\sin(\theta)$ (λ=neutron wavelength, θ=scattering angle) the repeating distance d can easily be calculated. For a fully hydrated DMPC bilayer the d-spacing lies in the order of 63 Ångstroms [55]. While the influence of the hydration on the structural properties of the membranes is well investigated, the knowledge about this influence on the lipid dynamics of these model systems is rather poor. From quasi-elastic neutron scattering investigations on pure lipid multilayer systems, the diffusive motions of lipids within the bilayer have been estimated. Consistent values for the diffusion constant in the order of $D=11*10^{-10}$ m^2/s [51-53] have been obtained. However, none of them take into account hydration effects.

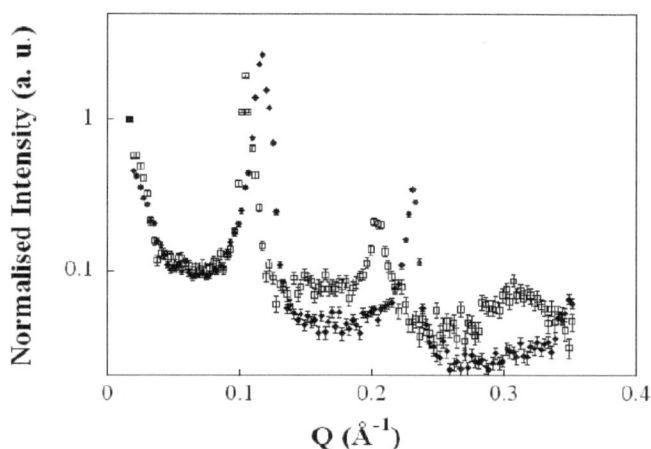

Figure 8: Diffraction data taken on D16 at ILL to evaluate the d-spacing for two samples with different hydrations levels [56]. Open squares: 100 % rh (pure D_2O atmosphere); filled diamonds: 75% rh (D_2O in saturated NaCl salt solution). The experimental d-spacing are 62.5 Å and 54.9 Å, respectively [56].

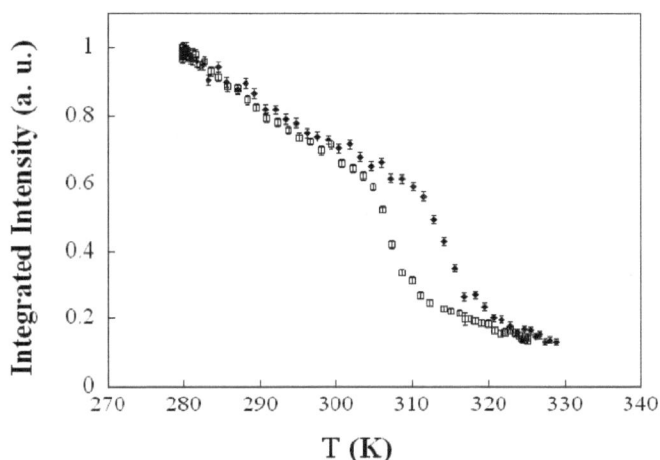

Figure 9: Integrated elastic intensity of DMPC, measured at 135°, on the cold neutron backscattering spectrometer IN16 at ILL with an energy resolution of 0.9 μeV. Data are binned over the Q range 0.43 – 1.93 Å^{-1} to gain in statistic. Empty symbols: h=100 % rh; filled symbols: h=75 % rh [56].

Only recently, a neutron scattering investigation, performed on highly oriented DMPC multibilayers, at two different rh levels, enhanced the strong influence of hydration effect on the membrane dynamics [56] (Fig. **9**).

The investigated temperature range covers both the main phase transition from the P_β ripple to the liquid-crystalline L_α phase which occurs around 296K for DMPC, at the hydration here investigated, and also the pre-transition from the L_β gel phase to the P_β ripple phase about 10 degrees below the main phase transition. The higher hydrated sample (100 % rh) shows a clear bend at the main lipid phase transition around 300K, in agreement with an increased mobility of the alkyl chain in the liquid phase, whereas a clear shift of the phase transition temperatures for the less hydrated sample (75 % rh) is evident.

The study reveals the strong influence of the hydration, not only on the structure but also on the dynamics of membrane systems.

REFERENCES

[1]   Henkel T, Mittler S, Pfeiffer W, Rotzer H, Apell HJ, Knoll W. Lateral order in mixed lipid bilayers and its influence on ion translocation by gramicidin: a model for the structure-function relationship in membranes. Biochimie 1989; 71: 89-98.
[2]   Pfeiffer W, Henkel TH, Sackmann E, Knoll W, Richter D. Local dynamics of lipid bilayers studied by Incoherent Quasi-Elastic Neutron Scattering. Europhys Lett 1989; 8: 201-6.
[3]   Deriu A. The power of quasielastic neutron scattering to probe biophysical systems. Phys Rev B 1993; 183: 331–42.
[4]   Doster W, Cusack S, Petry S. Dynamical transition of myoglobin revealed by inelastic neutron scattering. Nature 1989; 337: 754–6.
[5]   Zaccai, G. 2000. How soft is a protein? A protein dynamics force constant measured by neutron scattering. Science. 288: 1604–7.
[6]   Martenson RE. Myelin: Biology and Chemistry, Boca Raton, CRC Press, 1992.
[7]   Luzzati V, Mateu L. Order-disorder phenomena in myelinated nerve sheaths. I. A physical model and its parametrization: exact and approximate determination of the parameters. J Mol Biol 1990; 215: 373-84.
[8]   Smith R. The basic protein of CNS myelin: its structure and ligand binding. J Neurochem 1992; 59: 1589-608.
[9]   Riccio P, Masotti L, Cavatorta P, De Santis A, Juretic D, Bobba A, Pasquali-Ronchetti I, Quagliarello E. Myelin basic protein ability to organize lipid bilayers: structural transition in bilayers of lysophosphatidylcholine micelles. Biochem Biophys Res Commun 1986; 134: 313-9.
[10]   Carnegie PR, Dunkley PR, in: Agranoff BW, Aprison MH (Eds.), Advances in Neurochemistry, New York, Plenum Press, 1975.
[11]   Omlin FX, deH Webster F, Palkovits CG, Cohen S. Immunocytochemical localization of basic protein in major dense line regions of central and peripheral myelin. J Cell Biol 1982; 95: 242-8.
[12]   Boggs MJ. Lipid–protein interactions, Jost PC, Griffith OH, 1982, 2.
[13]   Inouye H, Karthigasan J, Kirschner DA. Membrane structure in isolated and intact myelins. Biophys J 1989; 56: 129-37.
[14]   Haas H, Torrielli M, Steitz R, Cavatorta P, Sorbi R, Fasano A, Riccio P, Gliozzi A. Myelin model membranes on solid substrates. Thin Solid Films 1998; 627: 327–9.
[15]   Facci P, Cavatorta P, Cristofolini L, Fontana MP, Fasano A, Riccio P. Kinetic and structural study of the interaction of myelin basic protein with dipalmitoylphosphatidylglycerol layers. Biophys J 2000; 78: 1413-9.
[16]   Natali F, Relini A, Gliozzi A, Rolandi R, Cavatorta P, Deriu A, Fasano A, Riccio P. Protein-membrane interaction: effect of myelin basic protein on the dynamics of oriented lipids. Chem Phys 2003; 292: 455-64.
[17]   Friedmann T. Overcoming obstacles to gene therapy. Sci Am 1997; 276: 96–101.
[18]   Rubanyi GM. The future of human gene therapy. Mol Aspects Med 2001; 22: 113–42.
[19]   Felgner PL, Gadek TR, Holm M, Roman R, Chan RW, Wenz M, Northrop JP, Ringold GM, Danielsen M. Lipofection: a highly efficient, lipid-mediated DNA-transfection procedure. Proc Natl Acad Sci 1987; 84: 7413–7.
[20]   Cullis PR, de Kruijff B. Lipid polymorphism and the functional roles of lipids in biological membranes. Biochim Biophys Acta 1979; 559: 399–420.
[21]   Farhood H, Serbina N, Huang L. The role of dioleoyl phosphatidylethanolamine in cationic liposome mediated gene transfer. Biochim Biophys Acta 1995; 1235: 289–95.
[22]   Liu Y, Mounkes LC, Liggitt HD, Brown CS, Solodin I, Heath TD, Debs RJ. Factors influencing the efficiency of cationic liposome-mediated intravenous gene delivery. Nat Biotechnol 1997; 15: 167–73.
[23]   Radler JO, Koltover I, Salditt T, Safinya CR. Structure of DNA-cationic liposome complexes: DNA intercalation in multilamellar membranes in distinct interhelical packing regimes. Science 1997; 275: 810– 4.

[24]   Crook K, Stevenson BJ, Dubouchet M, Porteous DJ. Inclusion of cholesterol in DOTAP transfection complexes increases the delivery of DNA to cells *in vitro* in the presence of serum. Gene Ther 1998; 5: 137–43.

[25]   Hirsch-Lerner D, Barenholz Y. Hydration of lipoplexes commonly used in gene delivery: follow-up by laurdan fluorescence changes and quantification by differential scanning calorimetry. Biochim Biophys Acta 1999; 1461: 47–57.

[26]   Kennedy MT, Pozharski EV, Rakhmanova VA, MacDonanld RC. Factors governing the assembly of cationic phospholipid-DNA complexes. Biophys J 2000; 78: 1620–33.

[27]   Regelin AE, Fankhaenel S., Gurtesch L., Prinz C., von Kiedrowski G., Massing. U. Biophysical and lipofection studies of DOTAP analogs. Biochim Biophys Acta 2000: 1464: 151–64.

[28]   Choosakoonkriang S, Wiethoff CM, Anchordoquy TJ, Koe GS, Smith JG, Middaugh CR. Infrared spectroscopic characterization of the interaction of cationic lipids with plasmid DNA. J Biol Chem 2001; 276: 8037–43.

[29]   Braun CS, Jas GS, Choosakoonkriang S, Koe GS, Smith JG, Middaugh CR. The structure of DNA within cationic lipid/DNA complexes. Biophys J 2003; 84: 1114–23.

[30]   Zuhorn IS, Oberle V, Visser WH, Engberts JBN, Bakowsky U, Polushkin E, Hoechkstra D. Phase behavior of cationic amphiphiles and their mixtures with helper lipid influences lipoplex shape, DNA translocation, and transfection efficiency. Biophys J 2002; 83: 2096–108.

[31]   Natali F, Pozzi D, Castellano C, Caracciolo G, Congiu Castellano A. DNA-lipoplex: a non-invasive method for gene delivery. A neutron scattering investigation. Biophys J 2005; 88(2):1081-90.

[32]   Castellano C, Natali F, Pozzi D, Caracciolo G, Congiu Castellano A. Dynamical properties of oriented lipid membranes studied by elastic incoherent neutron scattering. Physica B 2004; 350: 955-8.

[33]   Alberts B. Molecular Biology of the Cell., 3rd Ed; Garland Publishing Inc., New York, 1994.

[34]   Beitinger H, Vogel V, Möbius D, Rahmann H. Surface potentials and electric dipole moments of ganglioside and phospholipid bilayers: contribution of the polar headgroup at the water/lipid interface. Biochim Biophys Acta 1989; 984: 293–300.

[35]   Brocca P, Cantù L, Corti M, Del Favero E, Raudino A. Cooperative behavior of ganglioside molecules in model systems. Neurochem Res 2002; 27: 559–63.

[36]   Cantu' L, Corti M, Del Favero E, Muller E, Raudino A, Sonnino S. Thermal Hysteresis in Ganglioside Micelles Investigated by Differential Scanning Calorimetry and Light-Scattering. Langmuir 1999; 15: 4975–80.

[37]   Cantu' L, Corti M, Del Favero E, Raudino A. Tightly Packed Lipid Lamellae with Large Conformational Flexibility in the Interfacial Region May Exhibit Multiple Periodicity in Their Repeat Distance. A Theoretical Analysis and X-ray Verification. Langmuir 2000; 16: 8903–11.

[38]   Boretta M, Cantu' L, Corti M, Del Favero E. Cubic phases of gangliosides in water: Possible role of the conformational bistability of the headgroup. Physica A 1997; 236: 162–76.

[39]   Maggio B. The surface behavior of glycosphingolipids in biomembranes: a new frontier of molecular ecology. Prog Biophys Mol Biol 1994; 62: 55–117.

[40]   Simons K, Ikonen E. Functional rafts in cell membranes. Nature 1997; 387: 569–72.

[41]   Yuan C, Johnston LJ. Distribution of ganglioside GM1 in L-alpha-dipalmitoylphosphatidylcholine/cholesterol monolayers: a model for lipid rafts Biophys J 2000; 79: 2768–81.

[42]   Sharom FJ, Grant CW. A model for ganglioside behaviour in cell membranes. Biochim Biophys Acta 1978, 507, 280–93.

[43]   Peters M, Mehlhorn I, Barber K, Grant C. Evidence of a distribution difference between two gangliosides in bilayer membranes. Biochim Biophys Acta 1984; 778: 419–28.

[44]   Delmelle M, Dufrane SP, Brasseur R, Ruysschaert JM. Clustering of gangliosides in phospholipid bilayers. FEBS Lett 1980; 121: 11–4.

[45]   McIntosh TJ, Simon SA. Long- and short-range interactions between phospholipid/ganglioside GM1 bilayers. Biochemistry 1994; 33: 10477–86.

[46]   Bunow MR, Bunow B. Phase behavior of ganglioside-lecithin mixtures. Relation to dispersion of gangliosides in membranes. Biophys J 1979; 27: 325–37.

[47]   Sela BA, Bach D. Calorimetric studies on the interaction of gangliosides with phospholipids and myelin basic protein. Biochim Biophys Acta 1984, 771, 177–182.

[48]   Thompson TE, Allietta M, Brown RE, Johnson ML, Tillack TW. Organization of ganglioside GM1 in phosphatidylcholine bilayers. Biochim Biophys Acta 1985; 817: 229–37.

[49]   Natali F, Caronna C, Del Favero E, Deriu A, Cantu' L. Directional dynamics in DMPC membranes containing gangliosides. In preparation.

[50]   Smith GS, Sirota EB, Safinya CR, Clark NA. Structure of the L beta phases in a hydrated phosphatidylcholine multimembrane. Phys Rev Lett 1988; 60: 813-6.

[51]   Rheinstädter MC, Ollinger C, Fragneto G, Demmel F, Salditt T, Collective Dynamics of Lipid Membranes Studied by Inelastic Neutron Scattering, Phys Rev. Lett., 2004; 93: 108107-12.

[52]   König S, Pfeiffer W, Bayerl T, Richter D, Sackmann E. Molecular dynamics of lipid bilayers studied by incoherent quasi-elastic neutron scattering. J Phys II France 1992; 2: 1589-615.

[53]   Swenson J, Kargl F, Berntsen P, Svanberg C. Solvent and lipid dynamics of hydrated lipid bilayers by incoherent quasielastic neutron scattering. J Chem Phys 2008; 129: 045101-07.

[54]   Rheinstädter MC, Häußler W, Salditt T, Dispersion Relation of Lipid Membrane Shape Fluctuations by Neutron Spin-Echo Spectrometry, Phys. Rev. Lett., 2006; 97: 048103-09.

[55]   Kucerka N, Liu Y, Chu N, Petrache HI, Tristram-Nagle S, Nagle JF. Structure of fully hydrated fluid phase DMPC and DLPC lipid bilayers using X-ray scattering from oriented multilamellar arrays and from unilamellar vesicles. Biophys J 2005; 88: 2626-37.

[56]   Trapp M, Juranyi F, Unruh T, Tehei M, Gutberlet T, Peters J. To be submitted.

CHAPTER 4

Protein/Hydration Water Dynamics in Hard Confinement: Dielectric Relaxations and Picoseconds Hydrogen Fluctuations

Giorgio Schirò and Antonio Cupane[*]

University of Palermo, Dept. of Physics, Italy

Abstract: In this review we report on some experimental studies on the dynamics of Myoglobin in a confined geometry, obtained by encapsulation in a porous silica matrix, at low hydration levels. After formation through the sol-gel method, the samples were left aging/drying in order to reach a condition where only one or two water layers surround the proteins. In order to put in evidence the specific effect of confinement in the silica host, we compared this system with another one (*i.e.* hydrated powder) where proteins are confined by other proteins. Using elastic neutron scattering we investigate the temperature dependence of the mean square displacements of non-exchangeable hydrogen atoms of sol-gel encapsulated Myoglobin. In order to clarify the effect of hydration the study was extended to samples at 0.2, 0.3 and 0.5 [gr water]/[gr protein] fractions and comparison was made with Myoglobin powders at the same average hydration and with a dry powder sample. Comparison between the data relative to the different samples indicates that geometrical confinement within the matrix plays a crucial role in protein dynamics and conformational stability, the effect of sol-gel encapsulation being essentially a reduction of collective protein motions likely related to the slowing down of solvent confined diffusion. A dielectric spectroscopy investigation on the same systems helped us to clarify the effect of encapsulation on protein/solvent dynamics. In agreement with elastic neutron scattering, although in a much slower time scale, dielectric spectroscopy indicates a suppression of cooperative relaxation inside the gel, together with a clear dependence of relaxation rates on the hydration degree.

INTRODUCTION

The comprehension of the dynamical behavior of water and proteins in confined geometries and near solid surfaces is of utmost importance in biological physics, since most of the water in living organisms is closely associated with different kinds of biomolecules and/or with intracellular assemblies. In fact, it is well known that *in vivo* biological systems (e.g. proteins and other biomolecules inside the cell) are characterized by conditions of molecular crowding and geometrical confinement [1, 2]: these conditions could be extremely relevant in the comprehension of the relationship among dynamics, structure and function of proteins in their real biological environment.

Standard biophysical studies on protein solutions in a bulk phase and at low concentration may limit the understanding and neglect some key factors which determine structure, dynamics and also stability of biomolecules. An essential aspect of this complex scenario is therefore the role of confinement in restricted geometries and the influence of solid surfaces on the dynamics of globular proteins. In order to get a description of the effects of geometrical confinement on the dynamics of a protein-solvent system, we have chosen a suitable approach to obtain solid host matrices where we could confine inside nanometric cavities both protein and solvent through the sol-gel technique. Depending on sample treatment, it is possible to encapsulate proteins at different solvent composition (e.g. water-glycerol) and hydration level [3-9]. To highlight the specific effect of confinement in the silica host, we compared the encapsulated system with another one (*i.e.* hydrated powder, at the same average hydration level) where proteins are confined by other proteins. To study the dynamics of the encapsulated protein/solvent system, we used elastic incoherent neutron scattering (EINS) and broadband dielectric spectroscopy (BDS). EINS on D_2O-hydrated deuterated proteins explores mainly motions of the non-exchangeable hydrogen atoms of the protein and in the time scale of picoseconds to nanoseconds, depending on instrumental resolution.

Data on encapsulated Myoglobin (Mb) presented here reveal a clear effect of confinement, *i.e.* a reduction of anharmonic protein motions observed in sol-gel encapsulated samples, attributed to a perturbation of the collective dynamics (α-relaxation-like) of the solvent in the hydration shell due to confinement, more than to direct protein-matrix interactions. This effect depends non monotonically on the average protein hydration and presents a maximum at an

*Address correspondence to Antonio Cupane: University of Palermo, Dept. of Physical and Astronomical Sciences; E-mail: cupane@fisica.unipa.it

Salvatore Magazù and Federica Migliardo (Eds)

hydration level of about h=0.35 (hydration h defined as h=[g D_2O]/[g dry protein]). In view of these results, it is evident the importance of studying molecular relaxations of both proteins and solvent in confinement on a wider time scale.

Broadband dielectric spectroscopy (BDS) is an ideal technique to this purpose since it enables to extend the time scale of investigated motions from about 100 nanoseconds to about 100 seconds; moreover, being sensitive to relaxations of polar and charged molecular groups, it investigates the dynamics of water molecules and of protein side-chains. The temperature dependence of dielectric parameters (in particular of the relaxation time τ) allows to obtain a dynamic description of the processes involved in dielectric molecular rearrangements. In the second part of this review we present a set of BDS data on Mb encapsulated in silica hydrogel at two different levels of hydration, and in the temperature interval of 120-300 K. To study the effect of different confinements, also in this case the data are compared with those on Mb powders at the same hydrations. Aim of this study is to reveal the effect of confinement on dynamics on a wide time scale and to obtain a description of the coupling between solvent in the hydration shell and protein in the various environments.

MATERIALS AND METHODS

Samples

Lyophilised horse Mb (Sigma Aldrich) was dissolved in 0.1M K-phosphate buffer pH 7 (at room temperature). The ferric (met-Mb) state was guaranteed by the addition of a 4-fold molar excess of an oxidant agent (potassium ferricyanide, $C_6FeK_3N_6$); after equilibration, the excess potassium ferricyanide was removed by prolonged dialysis against a 0.1 M K-phosphate buffer solution at pH 7 at 7°C. The protein solution was then deuterated with several H_2O-D_2O (Euriso-Top, purity 99.97 %) exchanges *via* dilution/ concentration steps. The final D_2O percentage was greater than 99 %. The final protein concentration was 26.5 % in weight. Protein encapsulation in silica hydrogels was performed using the following protocol: a solution containing 60 % v/v TMOS (Sigma Aldrich), 38 % v/v D_2O and 2 % v/v HCl 0.04M was sonicated for 20 minutes in an ice bath; immediately after sonication it was mixed in a 1:1 proportion (in volume) and at 7°C with the met-Mb/D_2O solution. In these conditions formation of a gel about 1 mm thick occurred in about 1 minute. The gel was then left aging in a controlled atmosphere of N_2/D_2O and the hydration levels were determined from the observed mass change on drying (hydration h is here defined as h≡[gr D_2O]/[gr protein]).

H_2O encapsulated in silica gel was prepared using a procedure analogous to that described above: in the protocol the metMb/D_2O solution was substituted with pure H_2O.

Met-Mb powder was prepared with the following procedure: lyophilized horse Mb was dissolved in D_2O (Euriso-Top, purity 99.96 %) at a concentration of 50 mg/ml and held at room temperature for approximately one day. The solution was then centrifuged for 20 min at 10°C and subsequently lyophilised. The resulting lyophilised protein was held for about 30 hours under vacuum at 45°C and the obtained powder was considered our dry (h=0) sample. We are aware of the fact that the above procedure is unable to remove the strongly bound water molecules amounting at approximately 2 % w/w [10]; the hydration value relative to powder should therefore be considered as slightly overestimated. The powder was then held in atmosphere of D_2O and left to reach the desired hydration; hydration level was determined by measuring the mass change.

Elastic Neutron Scattering

Measurements

Elastic neutron scattering measurements as a function of temperature were performed on the thermal (λ=2.23Å) high-energy resolution backscattering spectrometer IN13 at the Institut Laue Langevin (Grenoble, France). IN13 allows to access the space and time windows of 1-6 Å and 0.1 ns respectively. In these experiments the energy resolution was fixed to 8 μeV. The elastic energy value (ω=0) was kept fixed within 3 μeV of accepted tolerance. The elastic scattering intensities S(Q, ω=0), suitably corrected for the empty sample holder contributions, were normalized with respect to the lowest temperature measurement to compensate for spurious background contributions and detector efficiency. In all the experiments, the sample thickness was suitably chosen to minimize the neutron absorption from the sample, thus avoiding correction from multiple scattering contributions. A transmission of about 88% was guaranteed using 1 mm thick Aluminium flat sample holder. No Bragg peaks due to crystalline heavy water were observed at all temperatures explored.

Data Analysis

The experimental data consist of the normalized incoherent elastic intensity $S(Q, \omega=0)$ at different temperatures. In the low Q region the Q-dependence of $S(Q, \omega=0)$ is given, in the frame of the Gaussian approximation by:

$$S(Q, \omega = 0) = I_0 e^{-\frac{\langle \Delta u^2 \rangle Q^2}{6}} \qquad (1)$$

where I_0 is a constant and

$$\langle \Delta u^2 \rangle \equiv \langle u^2(T) - u^2(20\,\mathrm{K}) \rangle \qquad (2)$$

is the normalized mean-square amplitude of the hydrogen atoms atomic displacements (MSD). The Gaussian approximation is valid for Q values that satisfy the condition $<\Delta u^2>Q^2< 2$ [11, 12]. However, as clearly shown from the data in Fig. **1**, for our samples, a deviation from a Gaussian behavior, particularly evident at high temperatures, is observed in the high Q region. This can be due to either extrinsic or intrinsic physical effects. Indeed in a protein the incoherent scattering mainly arises from hydrogen atoms, almost uniformly distributed in the entire molecule; the scattered intensity is therefore due to the superposition of scattering from all hydrogen nuclei, and it is a Gaussian only if it results from a summation of identical Gaussians. Either a summation of Gaussians with different $<\Delta u^2>$ (i.e. relative to potential wells of different widths) or a summation of identical non-Gaussian functions gives rise to a non-Gaussian form. The most plausible physical scenario is a coexistence of the two factors, but remarkable information can also be obtained by examining them separately. There are two proposed models taking into account this behavior.

Dynamical Heterogeneity Model

In the first model the non-gaussianity is attributed to the heterogeneity of hydrogen atoms MSD, resulting in a superposition of gaussian forms with different widths; a simplified version of this description has been introduced by considering a bimodal distribution [13-15]. The model assumes the validity of the Gaussian approximation but postulates the existence of a distribution function of hydrogen atoms MSD, $f(<\Delta u^2>)$, considered to be bimodal of the form:

$$f(\langle \Delta u^2 \rangle) = a_1 \delta(\langle \Delta u^2 \rangle_1) + a_2 \delta(\langle \Delta u^2 \rangle_2) \qquad (3)$$

where $\delta(x)$ is the delta function and a_1 and $a_2 = 1-a_1$ are the population fractions. Using Eq. (3) the scattering function becomes:

$$S(Q, \omega = 0) \propto \int_0^\infty f(\langle \Delta u^2 \rangle)\, e^{-\frac{\langle \Delta u^2 \rangle Q^2}{6}} =$$
$$= \int_0^\infty \left[a_1 \delta(\langle \Delta u^2 \rangle_1) + a_2 \delta(\langle \Delta u^2 \rangle_2) \right] e^{-\frac{\langle \Delta u^2 \rangle Q^2}{6}} = \qquad (4)$$
$$= a_1 e^{-\frac{\langle \Delta u^2 \rangle_1 Q^2}{6}} + a_2 e^{-\frac{\langle \Delta u^2 \rangle_2 Q^2}{6}}.$$

The choice of a simplified bimodal distribution is supported by several recent observations indicating that hydrogen dynamics in proteins arise from two main contributions, related to methyl group hydrogens and to all other non exchangeable hydrogens, respectively [14, 15].

Double-Well Model

In the second model, largely used in the literature, hydrogen atoms are considered all equivalent and can fluctuate in an anharmonic potential between two sites of different energy, whose minima (*i.e.* the equilibrium positions relative to each minimum) are separated by a distance d (asymmetric double-well model, originally proposed by Doster and co-workers [11]). The hydrogen atom can explore each site with a given probability (p_1 and p_2 respectively), with $p_2/p_1 = \exp(-\Delta G/RT)$, where ΔG is the free energy governing the process. In this context, the elastic scattering intensity is given by:

$$S(Q, \omega = 0) \propto e^{-\langle \Delta x^2 \rangle_G Q^2} \left\{ 1 - 2p_1 p_2 \left[1 - \frac{\sin(Qd)}{Qd} \right] \right\} \qquad (5)$$

where $\langle \Delta u^2 \rangle_G$ describes the Gaussian contribution to the mean square displacement. The second term represents the onset of new degrees of freedom of the system, resulting in the onset of an anharmonic dynamics. The total hydrogen mean square displacement is then given by:

$$\langle \Delta x^2 \rangle_{\text{tot}} = - \left(\frac{d \ln[S(Q, \omega = 0)]}{d(Q^2)} \right)_{Q=0} = \langle \Delta x^2 \rangle_G + \frac{p_1 p_2 d^2}{3} \qquad (6)$$

which is the initial slope of the curves of the logarithm of normalized scattering intensities vs. Q^2 at different temperatures. The term $\langle \Delta u^2 \rangle_G$, as discussed in ref. [11], is given by two Gaussian contributions arising from a vibrational process, dominating in the low temperature region, and from an α-relaxation, described as a slow process which may be resolved only in the high temperature region. The α-process in Mb is shown as having a non-Lorentzian shape, but it is well recovered by a Cole-Davidson function:

$$S(Q, \omega) = -A_1(Q) \frac{\Im\{(1 + i\omega\tau_0)^{-b}\}}{\omega}. \qquad (7)$$

Several theories have been applied to explain the origins of this α-relaxation process, including the Mode-Coupling theory [16], in which the process is seen as a collective effect, its shape arising as a result of non-linear coupling between density fluctuations. On the other hand, the second term in Eq. (6) is associated to the contribution arising from the β-process falling into the elastic window. In the framework of the Mode-Coupling theory, the β-process corresponds to local motions of atoms in the cage formed by their neighbors.

Dielectric Spectroscopy

Measurements

Dielectric spectroscopy measurements were performed in the frequency range 10^{-2}–10^7 Hz using a Novocontrol Alpha analyzer. Accuracy of dielectric spectra is $\approx 2\%$. Diameter and thickness of the sample holder were 28 and 1 mm, respectively. The surfaces of electrodes in contact with the sample were covered by an insulating layer to minimize dc conductivity. The sample was cooled from room temperature to 120 K at a rate of about 4 K/min and then spectra were measured from 120 K to 280 K at intervals of about 10 K, consecutively.

Data Analysis

All spectra were fitted by a combination of Havriliak-Negami functions describing each relaxation process, a term for direct conductivity and an additional term to take into account effects of electrode and interface polarization due to the inhomogeneous nature of the samples (Maxwell-Wagner type effects):

$$\epsilon_r(\omega) = \sum_j \frac{\Delta \epsilon_j}{[1 + (i2\pi\nu\tau_j)^\alpha]^\beta} + \epsilon_\infty + i\frac{\sigma}{2\pi\nu} + (A + iB)\nu^{-\lambda} \qquad (8)$$

$0 < \alpha \leq 1$ and $0 < \beta \leq 1$ are empirical shape parameters describing the width and the asymmetry of the loss peak, respectively but they have no a direct physical meaning; λ ($0 < \lambda < 1$) is a parameter describing the fractal character of the underlying processes [17].

RESULTS

Elastic Neutron Scattering

Comparison of non-Gaussian Models

The logarithm of normalized elastic intensities as a function of Q^2 relative to the sample at a representative hydration level (h=0.2) is reported in Fig. **1** for some selected temperatures. Data in panel (a) refer to encapsulated met-Mb and those in panel (b) to met-Mb powder.

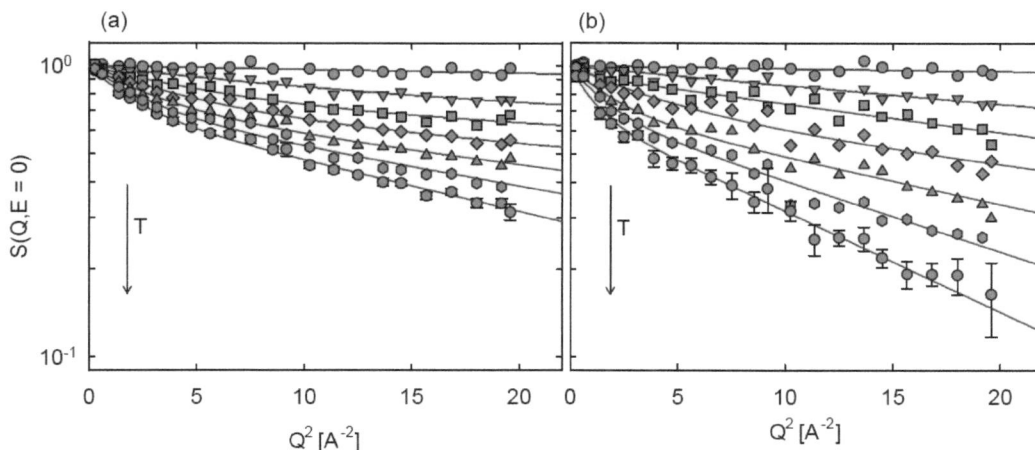

Figure 1: Normalized elastic intensities vs. Q^2 at various temperatures. Panel (a): met-Mb encapsulated in silica gel; panel (b): met-Mb hydrated powder. For the sake of clarity only few selected temperatures are drawn; from top to bottom: T = 50 K; T = 140 K; T = 190 K; T = 235 K; T = 260 K; T = 290 K; T = 316 K. The continuous lines represent fittings of the experimental data in terms of the dynamical heterogeneity model (cf. Eq. (4) in the text).

The main effect of confinement in silica gel, evident from a first inspection of data reported in Fig. **1**, is a reduction of dynamics of non-exchangeable hydrogens of met-Mb; this can be inferred from the much lower decrease of elastic intensity as a function of Q^2 for the encapsulated protein, mainly at high temperatures. The continuous lines are fittings of the experimental data obtained with the bimodal distribution (see Eq. (4)): parameters a1 and a2 have been set at 0.29 and 0.71 respectively at all the temperatures explored, since the fraction of methyl hydrogens over the total non-exchangeable hydrogens is 0.29 in horse Mb.

We note that elastic intensities are well fitted using this model over the entire Q and temperature ranges explored, thus indicating that a simplified bimodal distribution can take into account the main features of non-gaussianity in the Q-dependence of the elastic intensity. This result also supports the recent suggestion that two main contributions to the hydrogen dynamics in proteins are detectable, arising respectively from methyl group rotations and from motions of all other hydrogens [14]. It should be noted, however, that fittings of same quality (data not shown) are obtained also using the double-well model (see Eq. (5)) and that the data reported do not allow to distinguish between the two models. A behavior similar to that reported in Fig. **1** is observed at all the hydration levels investigated.

Fig. **2** shows the temperature dependence of the MSD obtained from the analysis with bimodal distribution for encapsulated met-Mb (black squares), in comparison with met-Mb hydrated powder (white circles). MSD relative to methyl and non-methyl hydrogens are reported in panel (a) and (b), respectively. Note that these MSD are plotted on an absolute scale, *i.e.* by adding to the 20 K-normalized MSD the contributions of zero point vibrations, assumed to be ≈ 0.01 Å^2 and ≈ 0.003 Å^2 for methyl and non-methyl hydrogens respectively. The continuous lines are fittings of the low temperature data (T < 130 K) in terms of a purely harmonic contribution given by

$$\langle \Delta x^2 \rangle = A \coth \left[\frac{h\nu}{2k_B T} \right] + C^2. \qquad (9)$$

Values of $A = 0.01$ Å^2 (0.002 Å^2 for non-methyl hydrogens), $\nu = 30$ cm-1 and $C2 \approx 0$ Å^2 are obtained from the fittings. It is remarkable that $\nu \approx 20\text{-}30$ cm^{-1} is the wavenumber of the so-called boson peak that dominates the neutron scattering spectra of proteins [18].

Data in Fig. **2** panel (a) show that, in agreement with data in ref. [14], two onsets of anharmonicity are observed in the dynamics of met-Mb. The first one, at about 100-150 K, involves methyl group rotations and its structural assignment is supported by similar observations in polymer systems at about the same temperature [19]. This assignment is directly proven by a recent EINS investigation on poly-alanine and poly-glycine hydrated powders; in fact, data in Fig. **3** show that both onsets, at ≈ 150K and at ≈ 230K respectively, are observed for poly-alanine (that has one methyl group in his side chain), while only the activation at ≈ 230K is observed for polyglycine (that has no methyl groups).

The second activation, at about 200-250 K, is the fingerprint of the well known dynamical transition and has been observed in the same temperature region in different biological molecules [20] with different techniques and can be put in relation with the activation of biological activity [21, 22]; it involves mainly larger scale motions and has been shown to be strongly coupled with solvent dynamics [20, 23-25].

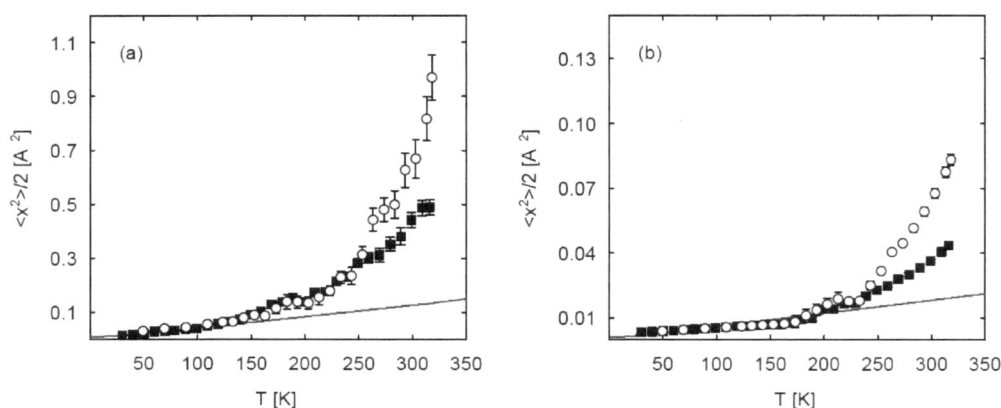

Figure 2: Temperature dependence of the MSD obtained with the dynamical heterogeneity model for the samples at hydration h=0.2. Black squares, met-Mb encapsulated in silica gel; white circles, met-Mb hydrated powder; continuous lines represent the harmonic contributions to the MSD. Panels (a) and (b): MSD of methyl and non-methyl hydrogens, respectively.

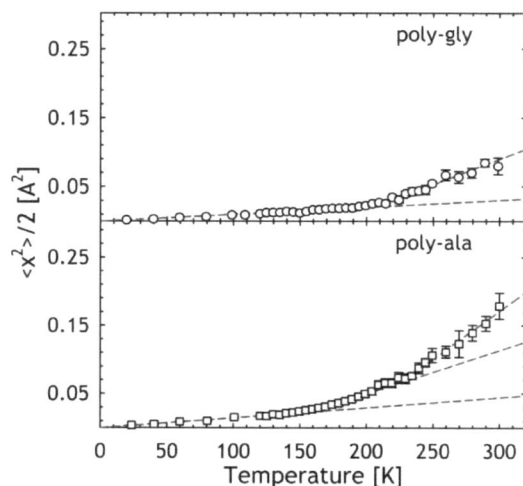

Figure 3: Mean square displacements of hydrogens in poly-glycine and poly-alanine as a function of temperature as determined from elastic scans on IN13. Hydration (D$_2$O) was h=0.2 w/w.

The temperature dependence of MSD relative to encapsulated met-Mb is identical (at least within experimental uncertainty) to that of met-Mb powder from 20 K up to about 230 K; at higher temperatures a drastic difference is observed and the MSD of met-Mb powder are systematically larger than those of encapsulated met-Mb.

This finding indicates that the effect of encapsulation on protein dynamics (as probed by the temperature dependence of mean square displacements) is mainly related to the so-called dynamical transition and involves a reduction of large scale slow motions (α-relaxation, in the terminology of ref. [11]). It can also be suggested that this effect is related to a modification of solvent dynamics: indeed, the internal dynamics of methyl groups is unaffected by encapsulation while the solvent-coupled dynamical activation above 230 K is clearly reduced in silica gel.

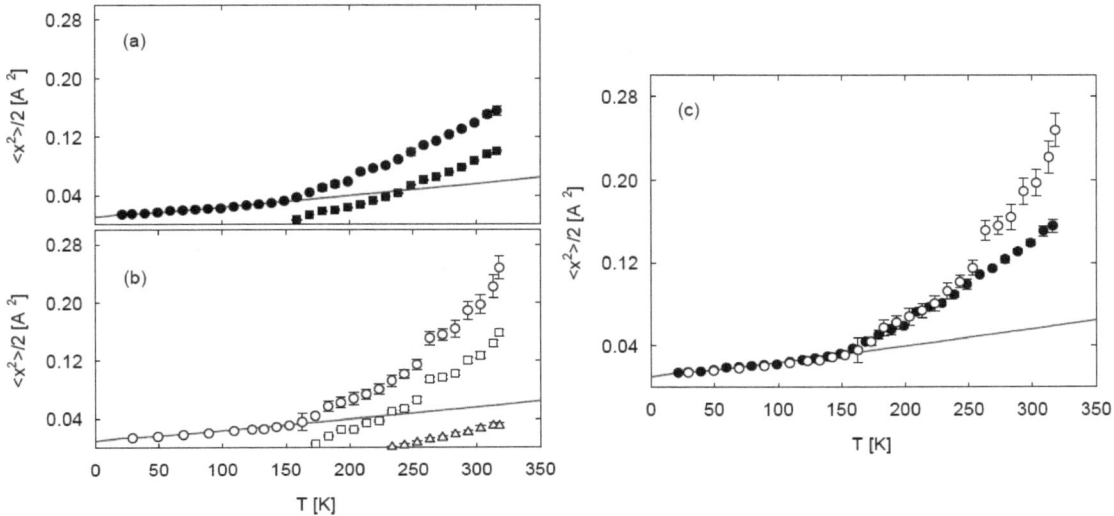

Figure 4: Temperature dependence of the MSD obtained with double-well model for the samples at hydration h = 0.2. Panel (a): met-Mb encapsulated in silica gel; black circles, total MSD; black squares, anharmonic contribution p1p2d2/3; continuous line represents the harmonic contributions to the MSD. Panel (b): met-Mb powder; white circles, total MSD; white squares, anharmonic contribution $p_1p_2d^2/3$; triangles, contribution of the α-relaxation process to the Gaussian MSD; continuous line represents the harmonic contributions to the MSD. Panel (c): comparison between met-Mb powder and met-Mb in silica ; black circles, total MSD for met-Mb in silica; white circles, total MSD for met-Mb powder.

The temperature dependence of the MSD obtained from an analysis in terms of the double-well model is reported in Fig. 4 (black symbols: encapsulated met-Mb; white symbols: met-Mb powder). Again, MSD values are reported on an absolute scale, assuming the zero point vibrations to be 0.014 Å^2 [26]. The continuous line is a fitting of the low temperature data (T < 130 K) in terms of a purely harmonic contribution (Eq. (9)) and it gives values of A = 0.003 Å^2, v = 30cm^{-1} and C2 = 0.009 Å^2. Values of the distance between the minima of the wells are d = 1.7±0.1 Å for both samples and are temperature independent. For the met-Mb powder sample (Fig. **4** panel (b)), in close analogy with the data by Doster *et al.*, 1989, a further contribution, besides the two described by Eq. 6, appears to be indispensable to fit the data and is reported by the white triangles symbols. This term is an additional Gaussian contribution (cf. Eq. (6)) to the pure vibrational term and it becomes observable at T > 250 K, fingerprint of an appearing new degree of freedom of the protein. However, this term is totally absent in the encapsulated met-Mb. The comparison of total MSD in Fig. **4** panel (c) confirms that encapsulation effects on mean square displacements become evident only at temperatures higher than about 200-250 K. The central result arising from the data reported in this section, *i.e.* that the large scale slow motions (α-relaxation) of protein are hindered in sol-gel encapsulated met-Mb, is clearly evidenced by both types of data analysis (i.e conformational heterogeneity model and double well model) and is therefore largely model independent.

Hydration Effects

In this section we discuss the effect of hydration on the dynamics of met-Mb in confined geometry. The study is here extended to samples at h = 0.2, 0.3 and 0.5 and comparison is made with met-Mb powders at the same average hydrations and with a dry powder sample. Since in the previous section it has been shown that main effects of encapsulation revealed by elastic data are model independent, for this comparison we report only analysis in terms of the dynamical heterogeneity model. MSD are reported in Fig. **5** as a function of temperature for met-Mb in silica hydrogel (right panels) and for met-Mb powder (left panels); data relative to a dry powder sample are also reported (stars in Fig. **5**), for a comparison. Data relative to methyl and non-methyl hydrogens are reported in the upper and lower panels, respectively. At all hydrations and for both the powder and silica systems, the two activations of anharmonicity previously discussed are observed. The second one, occurring at about 200-250 K, strongly depends upon hydration and is not observed in the dry powder sample. The fact that the dynamical transition in hydrated protein powders is solvent coupled, being absent in dry samples and approaching saturation at high hydrations, was already well established [11]; conversely, it is a new and intriguing result that a strong dependence of MSD behavior on hydration level is observed also in the encapsulated samples (Fig. **5**, panels (c) and (d)). This suggests that the effect of confinement on protein motions is mediated by the hydration shell.

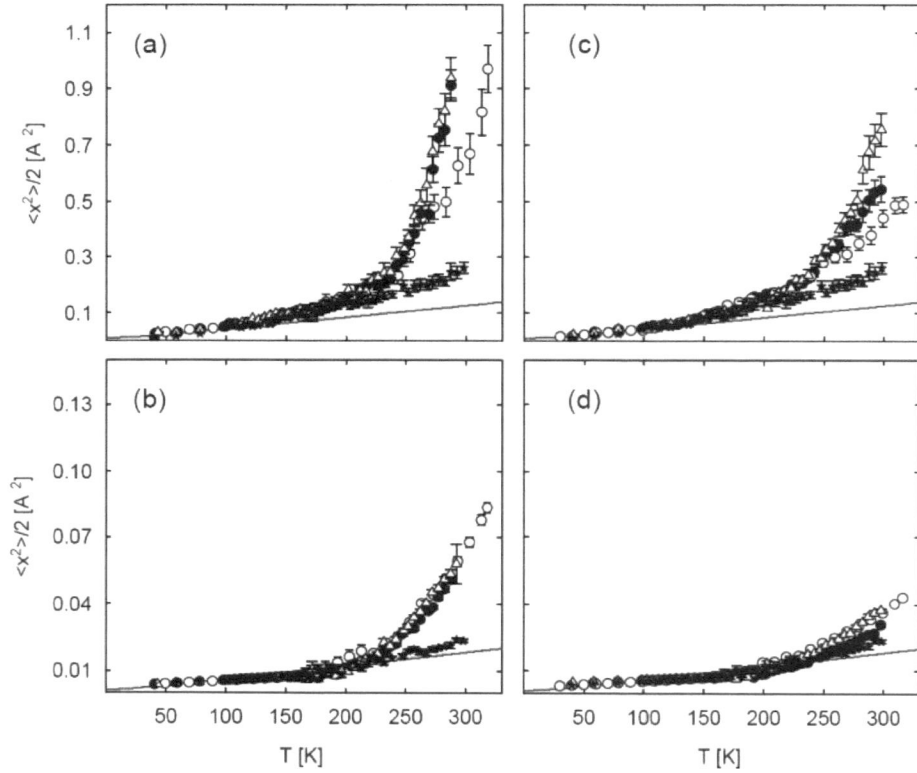

Figure 5: Temperature dependence of the MSD obtained using the dynamical heterogeneity model. Panels (a) and (b): MSD in met-Mb powder for methyl and non-methyl hydrogens respectively. Panels (c) and (d): same as panels (a) and (b), for met-Mb in silica gel. stars, h=0; white circles, h=0.2; black circles, h=0.3; triangles, h=0.5. Continuous lines represent the harmonic contributions to the MSD (see text).

A more detailed comparison between met-Mb in silica gel and met-Mb powder at different hydrations is presented in Fig. **6**.

It is evident that, at all the investigated hydration levels, the first onset of anharmonicity is unaffected by protein encapsulation while the solvent-coupled dynamical activation above 230 K is clearly reduced by protein encapsulation in silica gel. The confinement effect brought about by sol-gel encapsulation on protein dynamics can be quantified by defining the quantity:

$$\mathcal{F} = \frac{\langle \Delta x^2 \rangle_{\text{hydrated powder}} - \langle \Delta x^2 \rangle_{\text{hydrated gel}}}{\langle \Delta x^2 \rangle_{\text{hydrated powder}} - \langle \Delta x^2 \rangle_{\text{dry sample}}} \tag{10}$$

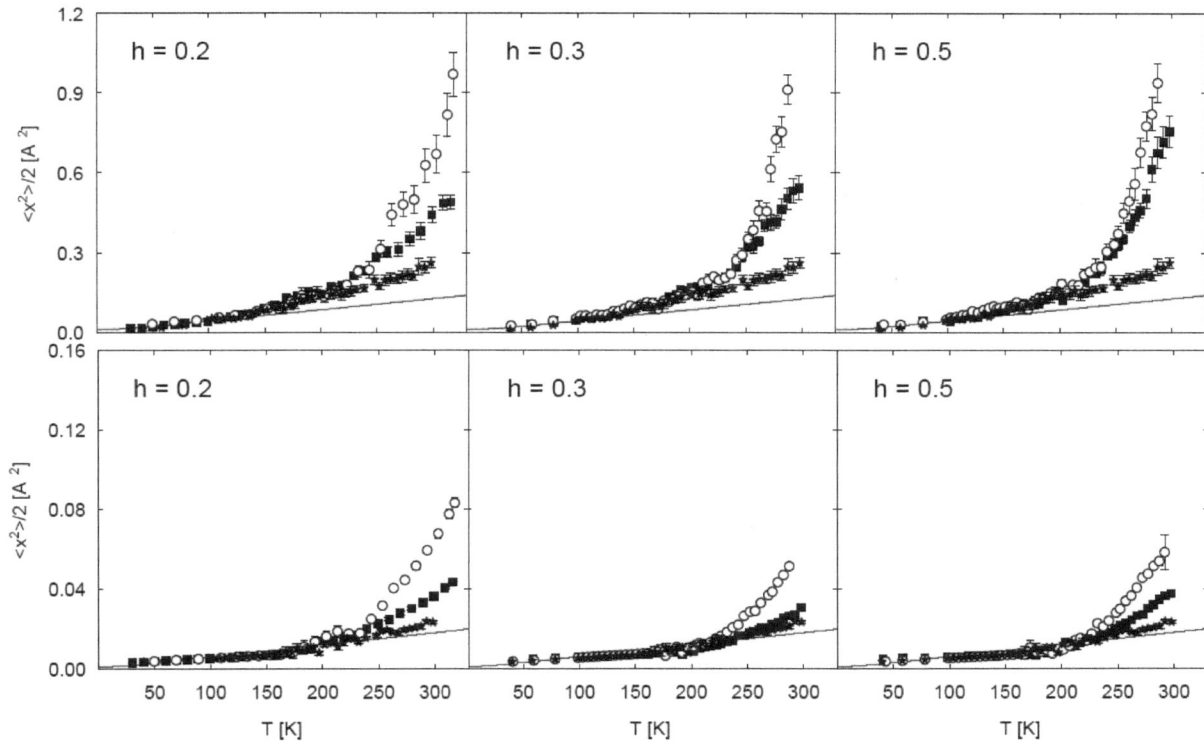

Figure 6: Comparison between MSD in met-Mb powder (white circles) and met-Mb in silica gel (black squares) at different hydration levels. Upper panels: methyl hydrogens contribution. Lower panels: non-methyl hydrogens contribution. Data relative to dry powder sample (stars) are superimposed for comparison. Harmonic contributions to the MSD are reported as continuous lines.

whose physical meaning is simply the fraction of large amplitude anharmonic motions that is absent in the sol-gel encapsulated sample due to the confinement effect. The quantity F calculated at T = 290 K is reported in Fig. **7** as a function of hydration (F shows the same dependence on hydration in the whole temperature range T > 250, but at higher temperature the relative error is smaller). As we can see, sol-gel encapsulation of met-Mb is able to hinder the anharmonic MSD of non-exchangeable hydrogens by more than 50 %; interestingly the stabilizing effect seems to have a maximum around 35 % hydration, *i.e.* at a value corresponding to a full hydration shell [27-29].

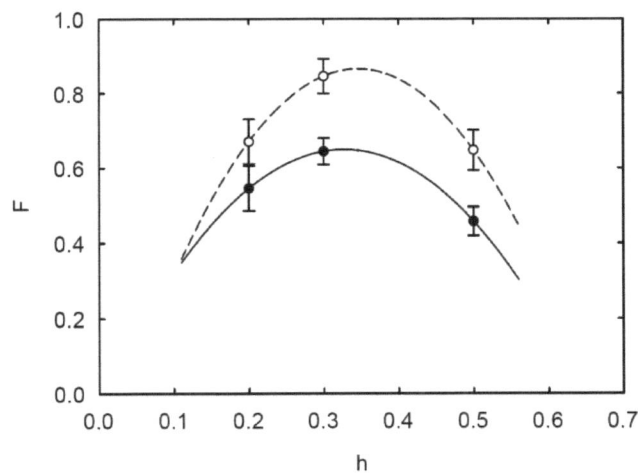

Figure 7: Fraction of confinement effect at T = 290 K as defined by Eq. 10 in the text, for methyl hydrogens (black circles) and non-methyl hydrogens (white circles). Lines are guides to the eye.

We can try now to rationalize the above effects in terms of protein-solvent interactions and of solvent dynamics in confinement. It has been proposed [24, 30-33] that a cooperative and properly structured hydrogen bond network is responsible for the correlation between protein conformational fluctuations and the dynamics of surrounding water molecules. Moreover, there are experimental evidences [34, 35] that water in hard confinement is characterized by a strong perturbation of hydrogen bond network, mainly due to the absence of a long-range bonding propagation, leading to partial (or even complete) disruption of cooperative water molecules motions and, depending on the hydration level, to the disappearance of the collective α-relaxation [36]. On the other hand, it has been proposed that large-scale, functionally relevant, protein motions are governed (slaved, in the terminology of ref. [23]) by the α-relaxation in the solvent. It is therefore reasonable to relate the reduction of the hydration dependent protein MSD observed at high temperature (T > 230 K) in sol-gel encapsulated met-Mb to a perturbation of solvent collective dynamics caused by confinement. Coherently, quasi-elastic neutron scattering data reported in ref. [7] showed that the diffusion coefficient of D_2O confined inside the pores of a silica matrix is reduced by at least one order of magnitude with respect to a bulk sample. The fact that the MSD reduction is more pronounced at hydration values near 35 % (see Fig. 7) suggests that solvent molecules in the first hydration shell are mostly involved in this effect.

We could wonder now why solvent dynamics (and, in turn, the solvent-coupled protein motions) should be so different in the two systems, silica gel and hydrated powder, since also in hydrated powder solvent is likely to be confined in a nanometer or sub-nanometer geometry. A tentative explanation could consider the recent teraHertz spectroscopy data suggesting that the biological relevance of water-protein interaction resides in the coherence between correlated water motions and protein fluctuations [37, 38]. Within this scenario, we could think that while in hydrated powders the solvent shell around one protein is confined by other fluctuating protein/solvent molecules, inside the silica matrix solvent layers on the surface of met-Mb are confined by the rigid walls of the matrix pores that are not able to preserve the coherent fluctuations of the hydrogen-bonded solvent network.

Dielectric Relaxations

Relaxations in Met-Mb Hydrated Powders

In Fig. **8** three-dimensional plots of dielectric losses measured in met-Mb powder are reported (left panel: h = 0.3; right panel: h = 0.5).

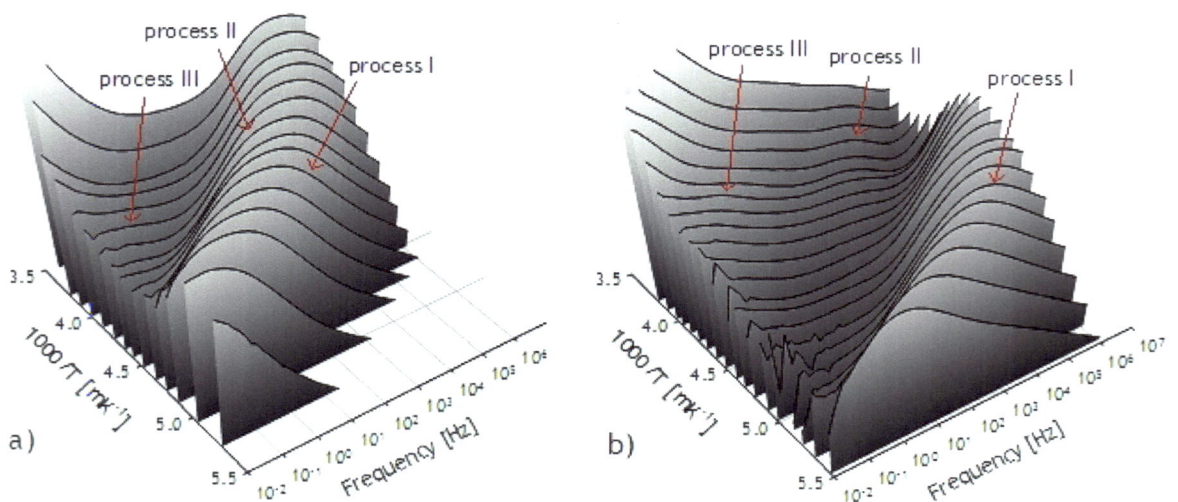

Figure 8: Dielectric losses in met-Mb powder at two different hydrations: h = 0.3 (left panel) and h = 0.5 (right panel).

Fig. **9** reports two typical dielectric spectra (imaginary part) at selected temperatures and relative to the two different hydrations investigated (left panels, (a): h = 0.3; (b): h = 0.5), and the respective Arrhenius plots of relaxation times (right panels, (c): h = 0.3; (d): h = 0.5). It is evident that the relaxation dynamics in the powder system is quite complex and several dielectric processes are present. In both samples a main relaxation process (process I) is observed (white circles and white squares in Figs. **9c** and **9d**, respectively), which shows a temperature dependence described by a Vogel-Fulcher-Tamman (VFT) behavior. This is the fastest process, and the relaxation times depend on hydration level: in fact, there is a difference of about three orders of magnitude in time between the two samples. These relaxations are described by a Havriliak-Negami function, with an evident asymmetry.

A second relaxation process (process II) is present (corresponding to black circles and black squares in Figs. **9c** and **9d**, respectively) with an interesting behavior as a function of temperature: in both samples the relaxation times appear to follow a VFT dependence in the high temperature region (T > 230-240 K, as indicated by vertical dotted lines in the Figs. **9c** and **9d**), while they deviate from this trend at lower temperatures and show an Arrhenius dependence, thus indicating that the activation energy of the process becomes constant. Swenson and co-workers [35] individuated at least two relaxation processes in Mb at hydration h = 0.8, and assigned the fastest process to interfacial water dynamics, whereas the slower one is attributed to motions of polar side chains on the protein surface, possibly with contributions from tightly bound water molecules. We think that the same assignment is valid also for the two processes observed in our samples; further support comes from the following arguments.

a)   The marked dependence of relaxation times for process I (white circles and white squares in Figs. **9c** and **9d**) on hydration finds a straightforward explanation if this process concerns hydration water dynamics. Indeed, we have to consider that h ≈ 0.35 corresponds to a complete hydration layer around the protein [27, 28, 29]; at higher hydration levels (like h = 0.5) a number of water molecules doesn't interact directly with protein surface (in particular with polar groups): they move more freely [39, 40] and make faster the dynamics of first hydration layer. The attribution of the process to water molecules is supported also by the different relative amplitude of the first process for the different hydrations: indeed the amplitude is larger if the water content is higher.

b)   A further indication on the origin of process I comes from Differential Scanning Calorimetry (DSC) data reported in ref. [8]. Indeed, the glass transition measured by DSC is clearly attributable to hydration water; on the other hand, the "dielectric" glass transition temperatures, estimated as usual from dielectric data by calculating the temperatures T_{100} that correspond to a relaxation time of 10^2 s (see Fig. **9**), are largely compatible with the "calorimetric" glass transition temperature region found by DSC [8]. Since the glass transition arises mainly from the α-relaxation, the coincidence between "calorimetric" Tg and "dielectric" T_{100} suggests that the solvent relaxation responsible for process I mainly involves the collective α-like relaxation of water in the hydration shell of the protein.

c)   Hydrated protein powders have been already investigated by dielectric spectroscopy as a function of hydration level and these studies were reviewed in ref. [10]. In order to explain the origin of the main relaxation process, a model based on the percolation theory was advanced [10, 41, 42]. In particular, it was observed that there is a critical value of the hydration at which the capacitance sharply increases with increase in hydration level. The threshold h_c = 0.15 gr. of water per gr. of protein (referred to a sample of lysozyme) is independent of pH below pH = 9 and shows no solvent deuterium isotope effect. The fractional coverage of the protein surface at h_c is in close agreement with the prediction of theory for surface percolation. The protonic conduction process was interpreted as a percolative proton transfer along threads of hydrogen-bonded water molecules adsorbed on the protein surface, with ionizable groups as sources of migrating protons. A principal element of the percolation picture, which explains the invariance of h_c to change in pH and solvent, is the sudden appearance of long-range connectivity and infinite clusters at the threshold hc. In this model, an increase of relaxation rate with increasing hydration is expected, and, in fact, measured [41, 42]. This relaxation process was observed in the MHz frequency range (at room temperature and in our hydration range) and showed a VFT dependence on temperature [43]: then it can be compared with the fastest process observed in our samples. In any case, also in the percolation framework this process is attributed to the dynamics of hydration water.

d)   Concerning the crossover observed in the second process (black circles and black squares in Figs. **9c** and **9d**), it should be noted that it occurs in the same temperature region (200-250 K) where elastic neutron scattering reveals the activation of large scale protein motions (α-like relaxations in the framework of double-well model). Although concerning motions occurring in widely different time scales (picoseconds vs. microseconds or slower) the correlation between BDS and EINS data suggests that the dynamical transition observed in the temperature dependence of hydrogen atoms mean square displacements is effectively a crossover from a non cooperative dynamics toward a collective α–like relaxation process. As regards the similarity in the temperature dependencies of process II and process I at T > 250 K, this could be a fingerprint of dynamical coupling between protein and adsorbed water: in their x-ray structure refinement of human and tortoise egg white lysozyme, Blake *et al.* 1983 [44] compared the average displacements of water molecules with the displacements of the hydrogen-bonded protein ligands. Water molecules bound simultaneously to two ligands had thermal displacement factors comparable to those of the associated protein atoms, whereas those water molecules bound to only one ligand exhibited higher displacements. Therefore, the protein molecule and the water around it form a strongly coupled system. Two mechanisms appear to provide a qualitative understanding of this, namely mechanical damping of the protein motion by adsorbed water and a dynamic electrical coupling between the fluctuating electric dipoles of the adsorbed water and the polar side-groups of the protein molecule.

A third relaxation process (process III) is distinguishable in both met-Mb powder samples (dotted circles and dotted squares in Figs. **9c** and **9d**), much slower than the others and showing an Arrhenius temperature dependence (the absence of points at high temperatures for met-Mb at h = 0.5 is due to the high conductivity signal which makes difficult to clearly identify peak position). Relaxation times and activation energy are similar in the two samples: this can be explained by supposing that this relaxation is related to inner protein motions (*i.e.* to motions of side chains in the interior of the protein).

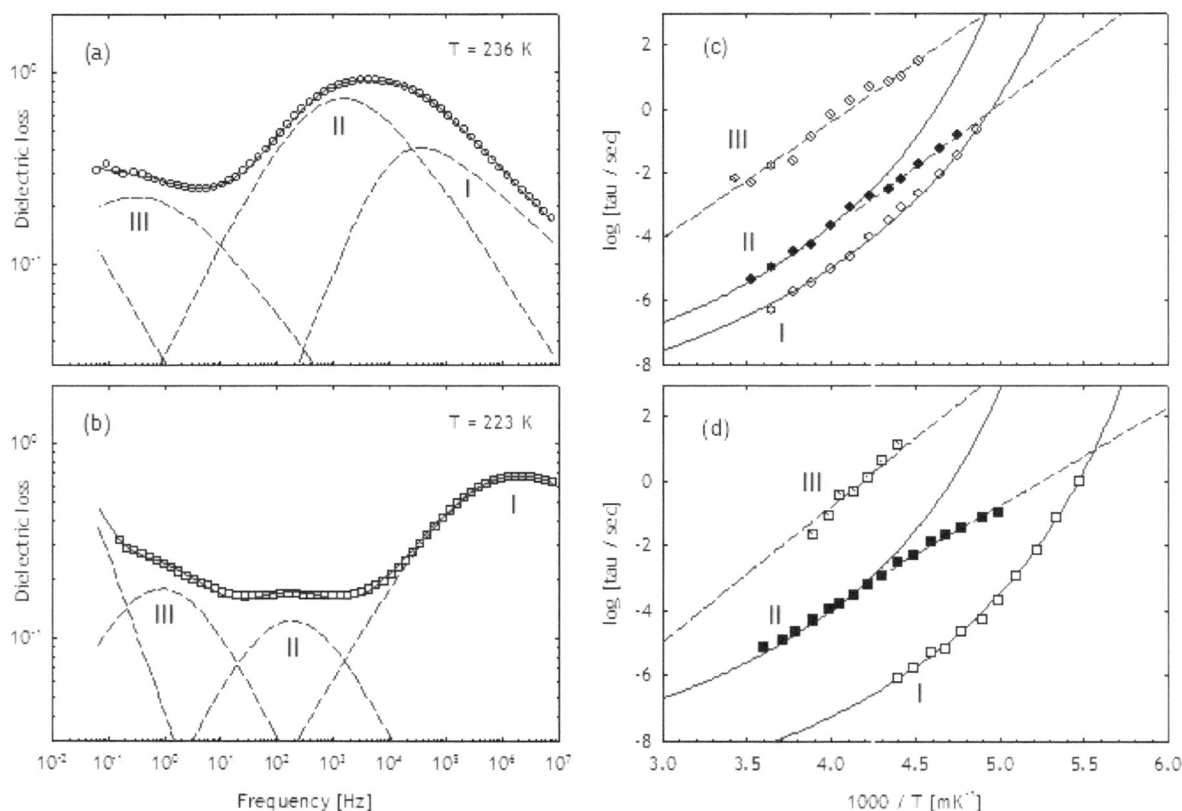

Figure 9: Dielectric relaxations in met-Mb powders at h=0.3 (upper panels) and h=0.5 (lower panels). Panels (a) and (b): typical dielectric spectra (imaginary part), at the temperature of 236 K (a) and 223 K (b); continuous lines are overall fittings of experimental data while dotted lines represent the individual processes. Panels (c) and (d): Arrhenius plots of relaxation times; continuous and dotted lines are fittings with VFT or Arrhenius law, respectively. The vertical thick line indicates the temperature of the VFT - Arrhenius crossover in the intermediate relaxation (process II).

Relaxations in Met-Mb Confined in Silica Gel

In Fig. **10** we show plots of dielectric losses measured in encapsulated met-Mb as a function of frequency and temperature (left panel: h=0.3; right panel: h=0.5). Fig. **11** reports typical spectra of dielectric loss as a function of frequency at a selected temperature, T=261 K (panel (a)) and the Arrhenius plot of relaxation times (panel (b)). The dielectric behavior of encapsulated samples is very different with respect to met-Mb powder system: as clearly shown in the figure, the dielectric response in the investigated frequency range (10^{-2}-10^7 Hz) is characterized by only one relaxation process in both samples; moreover, the temperature dependence of characteristic time, as obtained by a Havriliak-Negami fit (cf. Eq. 8) is well fitted by the Arrhenius law. The activation enthalpy, obtained from the slope of Arrhenius trend, is approximately the same for both samples, $\Delta H \approx 16$ kcal/mole. Note that the slope is also very similar to that of process III observed in the powder samples (see Fig. **12**). As a consequence we can argue that it is due to the same relaxation phenomenon, *i.e.* inner protein motions.

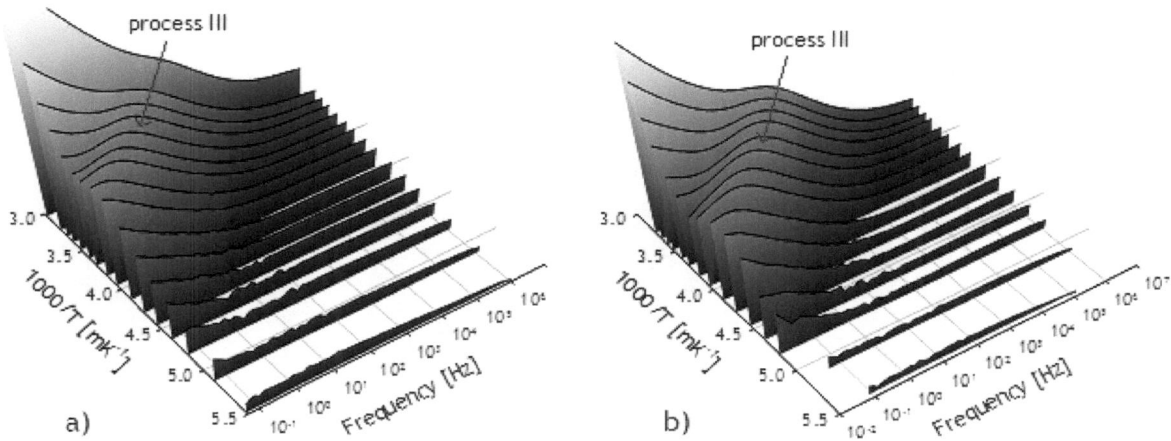

Figure 10: Dielectric losses in encapsulated met-Mb at two different hydrations: h=0.3 (left panel) and h=0.5 (right panel).

The most interesting finding arising from our data is that met-Mb confined in silica doesn't show (at least within the resolution of our measurements) the two faster relaxations (processes I and II) observed in the powder[1]. Since a non-Arrhenius character of the temperature dependence of a relaxation is suggestive of an underlying cooperative process, the absence of such relaxations in gel is a direct evidence of the main effect of confinement in silica at low hydration, *i.e.* the hindering of collective processes of both solvent and protein.

Further considerations on the possible origin of the Arrhenius process observed in encapsulated met-Mb are here in order. First of all, it should be noted that relaxation times are too large to be assigned to some dynamical process of water molecules. Rates of dielectric relaxation in water, even in the case of crystalline phase or in confined geometries, have never been observed in the timescale of this process. Moreover, the order of magnitude of activation energy is very close to the energy barriers ($\Delta H \approx 7 \div 23$ kcal/mole) estimated for sidechain motions of polar amino acids [45]. These considerations support the attribution of the process to some protein relaxation. However, if we compare the two different hydration levels, we observe a clear effect of hydration on relaxation time. Indeed, while the activation energy is not affected by the change in the water content, the rates are slowed down with decreasing hydration: the lower the hydration, the larger the relaxation times. An analogous effect is less evident in the powder samples. Moreover, as evident in Fig. **12**, the slopes of Arrhenius trend are almost coincident with the slope relative to the local β-like relaxation observed for pure water encapsulated in similar silica hydrogels (although, as expected,

---

1   In view of the percolative interpretation of the fastest process in met-Mb powders discussed in the previous section, a possible origin of the disappearance of this relaxation could be the insulating nature of silica host, which should prevent the existence of a long-range connectivity of proton network. However, this explanation can't take care of the suppression of intermediate relaxation process.

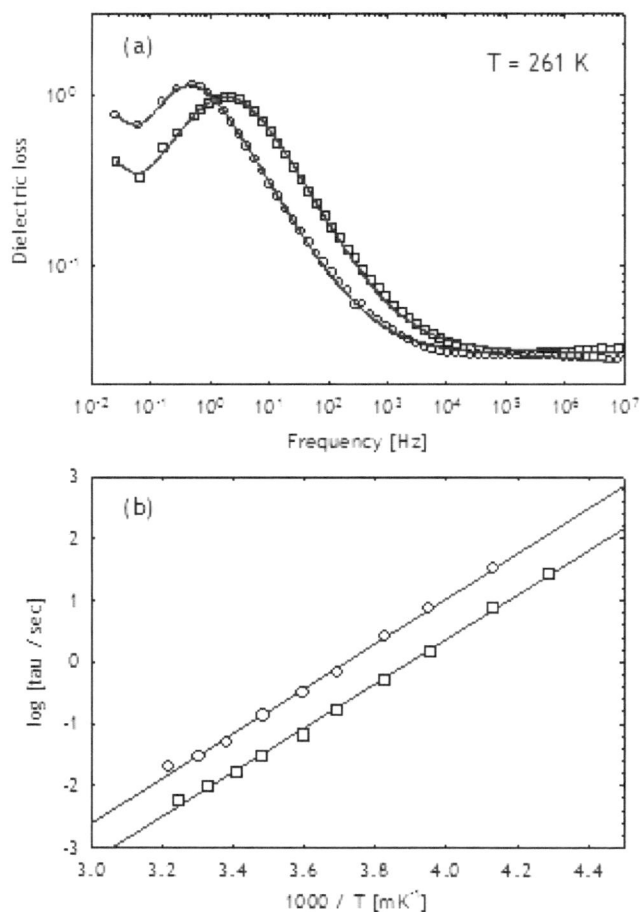

Figure 11: Panel (a): typical dielectric spectra (imaginary part) of encapsulated met-Mb at two different hydrations, at the temperature of 261 K; circles: h=0.3, squares: h=0.5. Panel (b): Arrhenius plot of relaxation times; symbols as in panel (a). Lines are fittings with Arrhenius law.

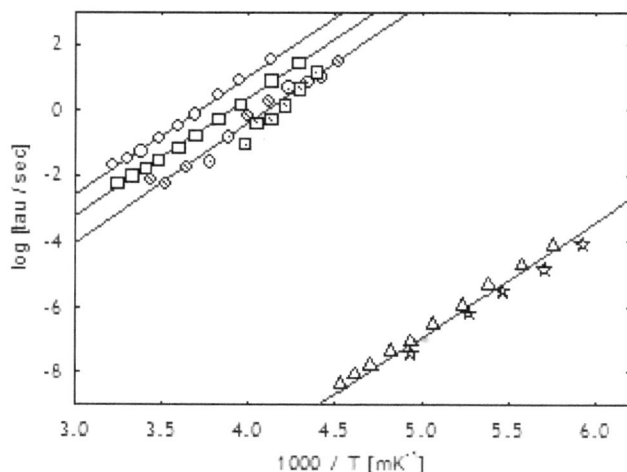

Figure 12: Temperature dependence of relaxation times relative to process III. Circles and squares: encapsulated met-Mb at h=0.3 and h=0.5, respectively; dotted circles and dotted squares: met-Mb powder at h=0.3 and h=0.5, respectively; stars and triangles: water encapsulated in silica gels (data from this work and from ref. [4]). Lines are fittings with Arrhenius law.

relaxation times for encapsulated water are much faster, see ref. [4] and next section). These effects are a clear manifestation of the "solvent slaving" effect described by Frauenfelder and co-workers [23]: β-like solvent relaxations induce local protein motions so that the activation enthalpy (*i.e.* the slope of the Arrhenius temperature dependence) is identical; on the other hand, actual relaxation times of protein processes remain much slower because of the protein + hydration shell acting as an entropy reservoir. A pioneering study [46] on dielectric relaxation of water adsorbed to crystals of met-Mb supports this scenario: the main result of that work is that the temperature dependence (*i.e.* the slope in the Arrhenius plot, not the timescale) of the GHz relaxation rate of adsorbed water is nearly identical to the temperature dependence of the conformational fluctuation rate of the protein as measured by the Mössbauer effect.

Comparison with a Protein-Free Silica Hydrogel

As shown in the previous sections, confinement in silica gel produces a specific effect on protein dynamics with respect to the hydrated powder, even at identical relative water content; moreover, protein dynamics is dependent on hydration level also in silica gel. The actual relevance of these findings is clearly related to the physical notion of hydration level and the possibility of comparing it in the two systems, *i.e.* silica gel and hydrated powder. This consideration suggests a deeper investigation about the connection between water content inside the gel and protein hydration. Do they coincide? Is the water inside the gel entirely around proteins? In order to address these questions, we prepared two new samples: the first one is met-Mb + H_2O (instead of D_2O) encapsulated in silica gel, and the second one is pure H_2O encapsulated in silica gel (protein-free sample). Both samples were left to age until they reached a stable weight. In Fig. **13** we report Arrhenius plots of relaxation times for the two samples. It is evident that the dielectric behavior is completely different: while two well distinct processes are oserved for the protein-free sample, one following a VFT temperature dependence and the other compatible with the Arrhenius law, only one, much slower, process is observed in the met-Mb + water sample. This implies that water molecules in the two samples experience very different dynamics because they experience a different confinement. In the presence of proteins water layers are confined only in the interstices between protein and silica surfaces. The absence in this system of the two main relaxations observed in protein-free sample clearly indicates the absence of pores containing only water. In this sense the physical meaning of hydration level defined as [gr water]/[gr protein] is the same in silica encapsulated samples and in powders. The dielectric behavior of water confined in a silica hydrogel has been already observed by Cammarata *et al.* [4] in an analogous sample: with the support of SAXS measurements these authors assigned the two relaxations to water molecules interacting with inner surfaces of silica pores and water molecules at the center of the pores and moving collectively. For a more extensive discussion about this attribution, see ref. [4].

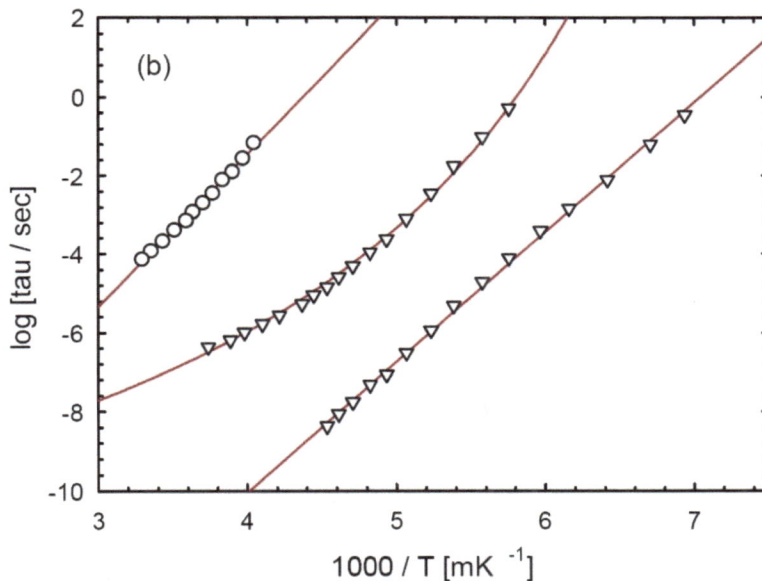

Figure 13: Comparison of dielectric relaxations in a silica gel containing met-Mb (white circles) and in a silica gel containing only water (triangles).

Our findings are in full agreement with those of Cammarata *et al.*; they support the picture of silica gel structure in the presence of template molecules arising from measurements performed through N_2 desorption isotherms [47, 48]. A silica hydrogel prepared *via* a sol-gel process without template molecules is characterized by a very broad and featureless distribution of pore dimensions; conversely, when template molecules like polymers, proteins or sugars are added in the sol phase the distribution of pore diameters becomes narrow and defines a well selected typical dimension of inner cavities, on the order of 30÷40 Å. The chemical explanation of these results is that during sol-gel process silica matrix condenses around the template, in our case around the protein+hydration water system.

CONCLUSIONS

Data reported in this review contribute towards the understanding of the effects of "hard confinement" on the dynamics of protein-hydration water systems. Conditions of "hard confinement" are obtained by encapsulating, *via* the sol-gel approach, the protein and its hydration water within a solid silica matrix. Dynamic properties are investigated with Elastic Incoherent Neutron Scattering and with Broadband Dielectric Spectroscopy, *i.e.* in the time scale of hundreds of picoseconds (EINS) and ranging from tens of nanoseconds to seconds (BDS). Effects of hard confinement are highlighted by comparing the results with those obtained on hydrated protein powders, at the same hydration level. The role of hydration water is investigated by extending the experiments to several hydration levels in the range from 0.2 to 0.5 [gr water] / [gr protein].

The main results obtained from EINS data may be summarized as follows:

- the effect of confinement mainly concerns large amplitude protein motions. In fact, while local motions like e.g. methyl groups rotations are barely affected, anharmonic, large amplitude, protein motions (also referred as α-like relaxations, [11]) that become activated above the dynamical transition temperature ($\approx$ 230K) are largely reduced by encapsulation within the silica hydrogel;

- the confinement effect, *i.e.* the reduction of the above mentioned anharmonic protein motions observed in sol-gel encapsulated confined samples, depends on the average protein hydration: it is enhanced at the hydration level corresponding to a complete water layer around the protein. This finding strongly suggests that confinement effect is mediated by the solvent.

The reported data can be rationalized in terms of protein-solvent interactions and of solvent dynamics in confinement, and suggest that the confinement effect is related to a perturbation of solvent collective dynamics (α-like relaxation) due to confinement more than to direct protein-matrix interactions.

The principal results arising from the BDS data are as follows:

- dielectric behavior observed in met-Mb powder is characterized by three different relaxation processes: 1) a VFT-like process attributable to hydration water dynamics, 2) an intermediate process, mainly attributable to relaxations of polar/charged side groups on the protein surface, and showing a crossover between VFT and Arrhenius temperature dependence, at T$\approx$200-250 K, the same temperature range where neutron scattering indicates the activation of α-relaxations, 3) a slow Arrhenius process mainly related to sidechains motions in the protein interior;

- in the encapsulated met-Mb sample the two cooperative processes observed in the powders are completely hindered and only a slow Arrhenius relaxation process is observable (corresponding to the process 3 in the powder cited in the previous item); it shows a dependence on hydration level that can be interpreted in terms of slaving model of protein dynamics by Frauenfelder and co-workers.

In conclusion, EINS and BDS data converge to a common picture of the hard confinement effect on protein dynamics, consisting mainly in the hindering of coherent/cooperative solvent motions in the protein hydration shell and, in turn, of solvent-coupled protein dynamics [37, 38]; at difference, local motions are largely unaffected. The close agreement is remarkable, given the widely different physical phenomena investigated by the two techniques (picosecond dynamics of non-exchangeable protein hydrogen atoms, EINS; water/protein side chains rotational/translational relaxations, BDS).

ACKNOWLEDGEMENTS

This work was supported by MIUR (Grant PRIN 2008ZWHZJT Struttura-Dinamica-Funzione di Biomolecole in Sistemi lontani dall'Idealità termodinamica) and local funds (ex-60%)

REFERENCES

[1]   Minton AP. The influence of macromolecular crowding and macromolecular confinement on biochemical reactions in physiological media. J Biol Chem 2001; 276: 10577-80.

[2]   Ellis RJ and Minton AP. Join the crowd. Nature 2003; 425: 27-8.

[3]   Levantino M, Cupane A and Zimanyi L. Quaternary structure dependence of kinetic hole burning and conformational substates interconversion in hemoglobin. Biochemistry 2003; 42: 4499-505.

[4]   Cammarata M, Levantino M, Cupane A, Longo A, Martorana A and Bruni F. Structure and dynamics of water confined in silica hydrogels: X-ray scattering and dielectric spectroscopy studies. Eur Phys J E 2003; 12: s63-6.

[5]   Schirò G and Cupane A. Quaternary relaxations in sol-gel encapsulated hemoglobin studied *via* NIR and UV spectroscopy. Biochemistry 2007; 46: 11568-76.

[6]   Schirò G, Cammarata M, Levantino M and Cupane A. Spectroscopic markers of the T → R quaternary transition in human hemoglobin. Biophys Chem 2005; 114: 27-33.

[7]   Schirò G, Sclafani M, Caronna C, Natali F, Plazanet M and Cupane A. Dynamics of myoglobin in confinement: An elastic and quasi-elastic neutron scattering study. Chem Phys 2008; 345:259-66.

[8]   Schirò G, Cupane A, Vitrano E and Bruni F. Dielectric relaxations in confined hydrated myoglobin. J Phys Chem B 2009; 113 : 9606-13.

[9]   Schirò G, Sclafani M, Natali F and Cupane A. Hydration dependent dynamics in sol-gel encapsulated myoglobin. Eur Biophys J 2008; 37:543-9.

[10]   Rupley JA and Careri G.. Protein hydration and function. Adv Protein Chem 1991; 41: 37-172.

[11]   Doster W, Cusack S and Petry W. Dynamical transition of myoglobin revealed by inelastic neutron scattering. Nature 1989; 337: 754-6.

[12]   Natali F, Ghiozzi A, Rolandi R, Relini A, Cavatorta P, Deriu A, Riccio P and Fasano A. Myelin basic protein reduces molecular motions in dmpa, an elastic neutron scattering study. Physica B 2000; 301: 145-9.

[13]   Nakagawa H, Kamikubo H, Tsukushi I, Kanaya T and Kataoka M. Protein dynamical heterogeneity derived from neutron incoherent elastic scattering. J Phys Soc Japan 2004; 73: 491-4.

[14]   Roh JH, Novikov VN, Gregory RB, Curtis JE, Chowdhuri Z and Sokolov AP. Onsets of anharmonicity in protein dynamics. Phys Rev Lett 2005; 95: 038101-9.

[15]   Doster W and Settles M. Protein-water displacement distributions. Biochim Biophys Acta 2005; 1749: 173-86.

[16]   Göetze W and Sjögren L. Relaxation processes in supercooled liquids. Rep Prog Phys 1992; 55: 241-370.

[17]   Liu SH. Fractal model for the ac response of a rough interface. Phys Rev Lett 1985; 55: 529-32.

[18]   Bizzarri AR, Paciaroni A, Arcangeli C and Cannistraro S. Low-frequency vibrational modes in proteins: a neutron scattering investigation. Eur Biophys J 2001; 30: 443-9.

[19]   Colmenero J, Moreno AJ and Alegria A. Neutron scattering investigations on methyl group dynamics in polymers. Prog Polymer Sci 2005; 30: 1147-84.

[20]   Caliskan G., Briber RM, Thirumalai D, Garcia-Sakai V, Woodson SA, Sokolov AP. Dynamic transition in tRNA is solvent induced. J Am Chem Soc 2006; 128: 32-3.

[21]   Rasmussen BF, Stock AM, Ringe D and Petsko GA. Crystalline ribonuclease a loses function below the dynamical transition at 220 K. Nature 1992; 357: 423-4.

[22]   Frauenfelder H, Sligar SG, Wolynes PG. The energy landscapes and motions of proteins. Science 1991; 254: 1598-603.

[23]   Fenimore PW, Frauenfelder H, McMahon BH and Parak FG.. Slaving: Solvent fluctuations dominate protein dynamics and functions. Proc Natl Acad Sci USA 2003; 99: 16047-51.

[24]   Fenimore PW, Frauenfelder H, McMahon BH and Young RD. Bulk–solvent and hydration–shell fluctuations, similar to α– and β–fluctuations in glasses, control protein motions and functions. Proc Natl Acad Sci USA 2004; 14408-13.

[25]   Cornicchi E, Onori G and Paciaroni A. Picosecond–time–scale fluctuations of proteins in glassy matrices: the role of viscosity. Phys Rev Lett 2005; 95: 158104-11.

[26]   Doster W.. in: Neutron Scattering in Biology, Methods and Applications. Fitter J, Gutberlet T, Katsaras J (Eds.). Springer Verlag, Berlin, 2006.

[27]   Steinbach PJ and Brooks BR. Protein hydration elucidated by molecular dynamics simulation. Proc Natl Acad Sci USA

1993; 90: 9135-9.

[28] Lounnas V and Pettitt BM. A connected cluster of hydration around myoglobin: correlation between molecular dynamics simulations and experiment. Proteins 1994; 18: 133-47.

[29] Doster W, Bachleitner A, Dunau R, Hiebl M and Luscher E. Thermal properties of water in myoglobin crystals and solutions at subzero temperatures. Biophys J 1986; 50: 213-9.

[30] Iben IET, Braunstein D, Doster W, Frauenfelder H, Hong MK, Johnson JB, Luck P, Ormos S. Schulte A, Steinbach PJ, Xie AH and Young RD. Glassy behavior of a protein. Phys Rev Lett 1989; 62: 1916-9.

[31] Tarek M and Tobias DJ. Role of protein-water hydrogen bond dynamics in the protein dynamical transition. Phys Rev Lett 2002; 88: 138101-9.

[32] Cornicchi E, Marconi M, Onori G and Paciaroni A. Controlling the protein dynamical transition with sugar-based bioprotectant matrices: a neutron scattering study. Biophys J 2006; 91: 289-97.

[33] Giuffrida S, Cottone G and Cordone L. Role of solvent on protein-matrix coupling in MbCO embedded in water-saccharide systems: a Fourier transform infrared spectroscopy study. Biophys J 2006; 91: 968-80.

[34] Bergman R and Swenson J. Dynamics of supercooled water in confined geometry. Nature 2000;, 403: 283-6.

[35] Swenson J, Jansson H.and Bergman R. Relaxation processes in supercooled confined water and implications for protein dynamics. Phys Rev Lett 1986; 96: 247802-6.

[36] Cerveny S, Schwartz GA, Bergman R and Swenson J. Glass transition and relaxation processes in supercooled water. Phys Rev Lett 1986; 93: 245702-7.

[37] Heugen U, Schwaab G, Bründermann E, Heyden M, Yu X, Leitner DM and Havenith M. Solute-induced retardation of water dynamics probed directly by terahertz spectroscopy. Proc Natl Acad Sci USA 2006; 103: 12301-6.

[38] Ebbinghaus S, Kim SJ, Heyden M, Yu X, Heugen U, Gruebele M, Leitner, DM and Havenith M. An extended dynamical hydration shell around proteins. Proc Natl Acad Sci USA 2007; 104: 20749-52.

[39] Gottfried O, Edvards L and Wüthrich K. Protein hydration in aqueous solution. Science 1991; 254: 974-80.

[40] Pal SK, Peon J and Zewail AH. Biological water at the protein surface: Dynamical solvation probed directly with femtosecond resolution. Proc Natl Acad Sci USA 2002; 99: 1763-8.

[41] Careri G, Geraci M, Giansanti A and Rupley JA. Protonic conductivity of hydrated lysozyme powders at megahertz frequencies. Proc Natl Acad Sci USA 1985; 82: 5342-6.

[42] Careri G, Giansanti A and Rupley JA. Proton percolation on hydrated lysozyme powders. Proc Natl Acad Sci USA 1986; 83: 6810-4.

[43] Pizzitutti F and Bruni F. Glassy dynamics and enzymatic activity of lysozyme. Phys Rev E 2001; 52905-10

[44] Blake CC, Pulford WC and Artymiuk PJ. X-ray studies of water in crystals of lysozyme. J Mol Biol 1983; 167: 693-723.

[45] Gelin BR and Karplus M. Sidechain torsional potentials and motion of amino acids in proteins: bovine pancreatic Trypsin inhibitor. Proc Nat Acad Sci USA 1975; 72: 2002-6.

[46] Singh GP, Parak F, Hunklinger S and Dransfeld K. Role of adsorbed water in the dynamics of metmyoglobin. Phys Rev Lett 1981; 47: 685-8.

[47] Wei Y, Xu JG, Feng QW, Lin MD, Dong H, Zhang WJ and Wang C. A novel method for enzyme immobilization: Direct encapsulation of acid phosphatase in nanoporous silica host materials. J Nanosci Nanotech 2001; 1: 83-93.

[48] Wei Y, Xu JG, Feng QW, Lin MD, Dong H and Lin M. Encapsulation of enzymes in mesoporous host materials *via* the nonsurfactant-templated sol-gel process. Mater Lett 2000; 44: 6-11.

SECTION III

Extreme Environments and Bioprotection Mechanisms

Proteins in Amorphous Saccharides: Structural and Dynamical Insights on Bioprotection

Sergio Giuffrida[1], Grazia Cottone[1,2], Alessandro Longo[3] and Lorenzo Cordone[1,*]

[1]*Dip. di Scienze Fisiche ed Astronomiche, Università di Palermo, I-90123, Palermo, Italy;* [2]*School of Physics, University College Dublin, Dublin, Ireland and* [3]*CNR-ISMN, I-90146, Palermo, Italy*

Abstract: We report on experimental and simulative insights on saccharide-based bioprotection, obtained through the study of proteins embedded in amorphous saccharide matrices. The data presented come from a complementary set of techniques (FTIR, MD simulations and SAXS), which provides a description of the bioprotection mechanism from the atomistic to the macroscopic level. The results concur to draw a picture in which bioprotection by saccharides can be explained in terms of a tight anchorage of the protein surface to a stiff matrix, *via* extended hydrogen-bond networks, whose properties are defined by all its components, and are strongly dependent on the water content. In particular, they show how carbohydrates having similar hydrogen-bonding capabilities exhibit different efficiency in preserving biostructures.

INTRODUCTION

Biopreservation is a relevant topic, in particular for its implications in food industry, pharmaceutics and medicine; large efforts are currently addressed to understand the mechanisms regulating the biopreservation processes, both *in vivo* and *in vitro*. In particular, amorphous saccharide matrices were proven to be very efficient in protecting biostructures against adverse conditions such as drought or extreme temperatures [1].

Among sugars, trehalose (α-D-glucopyranosyl-α-D-glucopyranoside), a non-reducing disaccharide commonly found in organism able to overcome adverse environmental conditions, resulted the best stabilizer of biostructures [2]; several hypotheses have been proposed to explain the origin of its effectiveness. The Water Replacement hypothesis [3] proposes that trehalose stabilization occurs through the formation of hydrogen bonds (HB) between the disaccharide and the embedded biostructure, in the dry state. This hypothesis is considered as the most valuable to explain the bioprotection of membranes by trehalose [4-6]. The Water Entrapment hypothesis [7] proposes that trehalose, rather than binding directly to biomolecules, entraps the residual water at the interface by glass formation, thus preserving the native solvation. This hypothesis is an extrapolation of the thermodynamic data by Timasheff and coworkers [8], who suggested that, in solution, saccharides are preferentially excluded from the protein domain. It has also been proposed that high viscosity could be at the basis of the reduction of large-scale internal protein motions leading to loss of structure and denaturation [9]. In this respect, Green and Angell [10] suggested the peculiarity of trehalose to be related to its glass transition temperature, which is higher than for analogous disaccharides. Furthermore, the very strong trehalose-water interaction may play a role in determining the trehalose bioprotective action. Raman and neutron scattering experiments [11-13] revealed the disruption of the tetrahedral network of water molecules on addition of trehalose, with a consequent reordering of water molecules around the saccharide, which impairs ice formation and improves preservation.

Another approach [14], based on structural studies of binary trehalose-water systems, proposes that the polymorphism of trehalose both in the crystalline and amorphous states is at the basis of its effectiveness. The slow formation of dihydrate trehalose crystals, by water evaporation at high temperature and low moisture, would keep water molecules in the same hydrogen bonding network as in the solvated trehalose, capturing the residual water molecules without disrupting the native structure of the biological systems; further slow dehydration would produce anhydrous trehalose, hence protecting the biostructures by inhibiting translational motions, preserving the active molecular conformations. This process has bioprotective effects due to the existence of reversible paths among the different states, avoiding at the same time water crystallisation.

***Address correspondence to Lorenzo Cordone:** Dipartimento di Scienze Fisiche ed Astronomiche, Università di Palermo, I-90123, Palermo, Italy; E-mail: cordone@fisica.unipa.it

Salvatore Magazù and Federica Migliardo (Eds)

The above hypotheses are not mutually exclusive. The formation of a glassy state does not imply hydrogen bonding, as evidenced by measurements on dextran [15]. The effeciency of trehalose may be due to the ability of forming glassy structures in a wide hydration range, along with the hydrogen bond capability. However, this is not consistent with results on raffinose, which is less effective than other sugars [16], although it has a glass transition temperature comparable with that of trehalose, along with larger hydrogen bonding potential [17].

It has been shown that the dynamics of a protein is highly inhibited when it is embedded in a trehalose glassy matrix [18-25], the inhibition being markedly dependent on the traces of residual water [21,25,26]. MD simulations [24,27,28] in protein-containing systems suggested that the protein surface is directly connected to a HB network including trehalose and water, in which the fraction of water molecules involved increases upon dehydration, so as the average number of hydrogen bonds in which each water molecule is involved.

The protein-matrix interaction, and its dependence on hydration, has been studied also by FTIR spectroscopy on carboxy-myoglobin (MbCO) embedded in trehalose matrices at different water content (water/sugar mole ratio from 0.3 to 3.0) [21,29,30]. The properties of the protein have been studied following the thermal evolution of the CO stretching band (COB, $\sim$1900–2000 cm^{-1}), which is split into three different sub-bands (A substates) [31], each corresponding to a specific different environment experienced by the bound CO within the heme pocket [32]. The relative intensity of these sub-bands depends on external parameters such as pH, temperature, and pressure [33]. The thermal behavior of the band therefore gives information on the thermal interconversion among taxonomic and lower hierarchy substates, which are evidenced, for example, by the thermal line broadening and peak frequency shifts. The properties of the matrix have been studied following the thermal evolution of the adjacent Water Association band (WAB, $\sim$2000-2400 cm^{-1}). This band is attributed to a combination of the bending mode of water molecules with intermolecular vibrational modes involving either other water molecules or non-water HB-forming groups [29,34]. Due to its intermolecular origin, the thermal evolution of this band was assumed to reflect the thermal rearrangements of the water-containing matrix. A rough quantitative evaluation of the thermal evolution of both the protein and the matrix was obtained through the, so called, Spectra Distance (SD) of the normalized spectra at various temperature, from the respective normalized spectrum measured at a reference temperature. This is defined as [29,35]:

$$SD = \left\{ \sum \left[A(v,T) - A(v,T_{ref}) \right]^2 \Delta v \right\}^{1/2} \tag{1}$$

where $A(v)$ is the normalized absorbance at the frequency v and Δv is the frequency resolution. The above quantity is the deviation of the normalized spectrum at temperature T from the normalized spectrum in condition of minimal occurrence of thermal motions, which, in the most of the cases, corresponds to the lowest temperature investigated (20 K). We propose it reflects the thermally induced readjustments of either the protein heme pocket (SD$_{CO}$) or of the matrix (SD$_{WAB}$).

Overall, FTIR and MD results suggested that the rigidity of the HB network increases on drying and that this network is mainly responsible for coupling the internal dynamics of the protein to that of the low-water matrix [21,36]. The inhibition of internal protein dynamics under extreme dryness can therefore be explained in terms of a tight anchorage of the protein surface to a stiff matrix: saccharide protect biomolecules not simply by preserving their native solvation, but rather by locking their surface through constrained water molecules, thus hindering motions. Accordingly, when a strong, extended HB network form, as in glassy trehalose–water–protein systems, the energy penalty associated with the solvent re-arrangements, needed for large-scale protein internal motions, increases sizably [37], as supported by vibrational echo experiments performed on different hemeproteins in trehalose glasses and silica gels [38].

In full agreement, it has been suggested [39,40] that water translational motions, which allow complete exchange of protein-bound water molecules by translational displacement, are necessary for large-scale fluctuations involving displacements of the protein surface [41], as e.g., interconversion among high tier substates and structural relaxations. Furthermore, it was also suggested [39,40] that exchange of protein-water HBs by water rotational/librational motions, not sufficient to permit large-scale internal motions, still allows interconversion among low tiers conformational substates, which does not involve displacements of the protein surface.

The formation of extended HB networks has been rationalized by Terahertz Spectroscopy (THz), an useful tool for probing the collective modes of the water network. In particular, it has been reported that each solute affects in a peculiar way the fast collective network motions of the solvent well beyond the first solvation layer and the total THz absorption depends on both saccharide concentration and number of HB formed with water [42]. This suggests that, under suitable solutes concentration, large-scale collective modes involving wide regions of the sample can set up, such long-range coherence failing if solute particles reduce their participation in hydrogen bonding.

In line with the above arguments, a Differential Scanning Calorimetry (DSC) study reported that, in myoglobin–trehalose–water systems below the water/disaccharide ratio at which the whole samples can form homogenous glasses, a linear correlation exists between the protein denaturation temperature T_{den} and the glass transition temperature T_g of the whole protein–trehalose–water system. This gave further information on the coupling between the protein stability and the dynamics of the surrounding matrix, indicating that the collective properties that regulate the glass transition are linearly correlated to local properties bound to the protein denaturation [43].

Notwithstanding the above reported data indicate a particular effectiveness of trehalose, also other sugars exhibit bioprotective properties and are used as bioprotectant by different organisms. Actually, this is to be expected due to the rather similar structure of the various saccharides, hence their similar hydrogen bonding capability. A deeper understanding of the trehalose peculiarity could be then reached only by comparing the properties of systems with different saccharides. In this short review we report on recent results on myoglobin embedded in some disaccharide matrices (maltose, sucrose, lactose, trehalose), as well as the trisaccharide raffinose. Studies on glucose pointed out that monosaccharides barely form hard amorphous matrixes, which are the most effective toward bioprotection. Different experimental techniques were used, along with Molecular Dynamics simulations, with the aim to obtain a picture of the bioprotection mechanisms from the atomistic to the macroscopic level.

INFRARED SPECTROSCOPY

Infrared Spectroscopy

Saccharides are very hydrophilic molecules, able to bind several water molecules in their hydration spheres. For this reason the properties of the solid amorphous saccharide matrices are strongly dependent on water content, and the evaluation, and control, of the hydration play an important role in the study of these matrices. This evaluation can be made with various absolute techniques, such as e.g. gravimetry, but the study of the properties of the infrared bands of water can convey deeper information on the hydration properties of these systems. As mentioned in the Introduction, the WAB arises from interaction of water molecules with either other water molecules or non-water HB-forming groups. Accordingly, two subpopulations of absorbers contribute to the band, making the dependence of its area on water content not linear. This non-linearity becomes relevant on drying, when a sizable fraction of the water molecules interacts with non-water HB-forming groups. At variance, an intramolecular band, such as the water combination band at 5200 cm^{-1}, ascribed to a combination of bending and asymmetric stretching [34], can be useful to have a less biased estimate of the sample water content, since only oscillator strength variations might affect the band during the drying. The ratio between the areas of the two bands, $r_A=A_A/A_C$, gives a rough estimate of the relative weight of water molecules interacting with non-water HB-forming groups in each sample, larger fractions of the latter corresponding to larger values [21]; r_A therefore constitutes a simple tool to evaluate the effective hydration of the system.

In table 1 we report the areas of the association and combination bands (A_A, A_C) and the ratio r_A for a selection of samples. We report on three level of hydration for the protein-containing samples and on only one for samples without protein. By comparing the water content for the samples without and with protein at the same hydration level (samples 0 and 1, respectively) it is evident that trehalose, maltose and raffinose samples without protein have similar r_A values, while for sucrose, because of the lower water content (indicated by the lower value of A_c), the r_A value is higher. The addition of protein causes a generalized increase of r_A values, due essentially to a reduction of residual water content, hence of the fraction of HBs used in the interaction with non-water hydrogen-bonding groups. This increase brings r_A for trehalose and maltose to a very similar value, while this value is slightly lower for sucrose and quite larger for raffinose. The sucrose behaviour can be rationalized by taking into account Molecular Dynamics results on binary sucrose/water systems [44], which showed that this sugar performs intramolecular hydrogen bonds at lower sugar/water ratio than trehalose. The more likely occurrence of internal HB would thus leave an overall lower number of sites available to HB with the surrounding, and this is reflected by the lower residual water content in the absence of protein and the lower r_A values in its presence [30]. At variance, in agreement with the large number of coordination water molecules present in its pentahydrate crystalline state [45], raffinose directly binds a larger fraction of water molecules than the other sugars [17], so in ternary systems a stronger competition for sugar is present, which concur to reduce the fraction of water-water interaction. This is also confirmed by the rather high r_A value for raffinose also in the following step of hydration (sample R2). In all the sugars, the increase in hydration, to type 2 and then 3, goes obviously with the reduction of r_A values, due to the larger availability of residual water. In hydrated type 3 samples water-water interactions are expected to dominate, as in all the saccharide matrices r_A value approaches the value in pure water (1.8) [30]. Unfortunately this can only be proven in ternary systems since corresponding binary systems are not stable enough on hydration and, for all the sugars studied, show a strong drift toward crystallisation [35].

Beside considerations on the band area, the WAB is primarily a marker of the HB network present in the matrix. Although a detailed study of the components of the band, which could shed light on the matrix structure, has not already been published (work in progress), the simple comparison of the band shapes in different matrices, with and without embedded protein, provides data on the effects the presence of the protein plays on the matrix structure. In Fig. **1** the normalized profiles of the water association band for four different saccharides (trehalose, sucrose, maltose and raffinose), in the presence and in the absence of the protein, are shown, all in the driest conditions obtained (types 1 and 0). We make use of normalized spectra since we are interested in changes of absorption profile. A similar study at higher hydration is not useful, as, beside crystallisation, the band quickly approaches the shape it has in pure water, losing all significant features [35]. WAB profiles in dry conditions are instead strongly dependent on the type of saccharide, as they arise from different pattern of population of HB local structures.

Table 1: Areas underlying the association band (A_A); areas underlying the combination band (A_C); ratio $r_A = A_A/A_C$, for a selection of samples. The values are reported from [29] for trehalose samples, [30] for ternary sucrose sample, [21] for ternary maltose and raffinose samples, [35] for binary sucrose, maltose and raffinose samples. All the samples were prepared with an analogous procedure, whose details are reported in the mentioned papers. It has been chosen to report on only one representative sample per type.

Sugar		A_A	A_C	r_A	Sugar		A_A	A_C	r_A
	T0	25.5	4.0	6.4		M0	31.1	5.3	5.9
Trehalose	T1	5.5	0.2	27.5	Maltose	M1	5.5	0.2	27.5
	T2	7.6	1.2	6.5		M2	19	5.3	3.6
	T3	31.5	18	1.8		M3	25.7	10.6	2.4
	S0	11.2	1.0	11.2		R0	26.3	4.6	5.7
Sucrose	S1	3.9	0.2	19.5	Raffinose	R1	13.6	0.2	68
	S2	9.6	3.3	2.9		R2	18.2	0.9	20
	S3	35.1	19	1.8		R3	29.3	11.4	2.6

The comparison of the band profile in binary and ternary systems show that the studied sugars fall in two categories: in trehalose and maltose (panels *a* and *c*) the band shape undergoes a huge change, indicating that the introduction of protein alters strongly the distribution of HB structures, while in sucrose and raffinose (panels *b* and *d*) the band shape show only minor changes. In particular in both maltose and trehalose one can notice an increase in the low frequency components, which originates from the presence of weaker hydrogen bonding groups, as expected in case of protein-water or protein-sugar interaction. This increase is just noticeable in sucrose and raffinose, indicating that the distribution of water molecules is barely altered, hence the protein has only a weak influence on the sugar matrix; conversely the matrix does not change its HB networks to better host the protein, and this can be at the basis of a reduced effectiveness toward bioprotection. An important outcome of these data is that the presence of the protein can have different effects on different environment, as a counterpart of the different effects on the protein structure, shown by different saccharide host media. It is not possible to infer if the changes induced by the protein are preferentially local to the protein surroundings or encompassing the whole sample basing on these data [46], although the protein-matrix coupling (see below) support the presence of a long range influence.

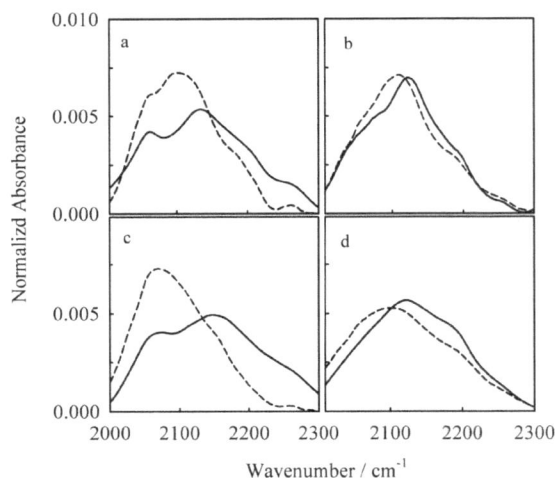

Figure 1: Normalized spectra of the WAB at 300 K for extra-dry samples, both with embedded MbCO (type 1, dashed lines) and without protein (type 0, solid lines), for the four studied saccharides: trehalose (panel *a*) [29], sucrose (panel *b*) [30, 46], maltose (panel *c*) and raffinose (panel *d*) [21].

The concurrent study of WAB and COB has been used to look at the protein alterations. In Fig. **2** the normalized CO stretching bands at 20 K in condition of extreme dryness (water/sugar mole ratio ~0.3) are shown, for samples of myoglobin embedded in matrices of trehalose, maltose, sucrose and raffinose. The bound CO stretching band is strongly dependent on the type of sugar as well as on the hydration of the samples [21,29,30].

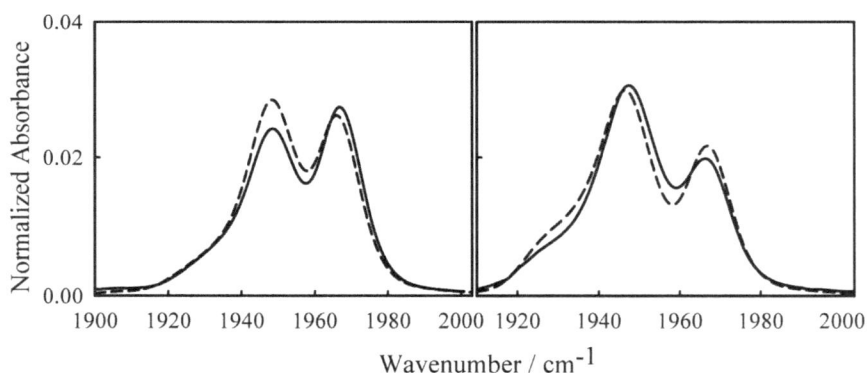

Figure 2: Normalized spectra of the COB at 300 K for extra-dry samples (type 1), for the four studied saccharides. Left panel: trehalose (solid) [29] and maltose (dashed) [21]. Right panel: sucrose (solid) [30] and raffinose (dashed) [21].

In particular, as evident from the figure, maltose and trehalose matrices have very similar CO spectra, differing only for a slightly lower population of the A_0 substate in the former, while in the low frequency side of the spectrum of sucrose and raffinose matrices a shoulder appears, which corresponds to a fourth A substate needed for the fitting of the band. This substate cannot be considered as the formerly reported A_2, at 1942 cm^{-1} [31], since its peak frequency is ~1925 cm^1 [30]. We suggest the occurrence of this fourth substate to arise from heme pocket structural modifications induced by sucrose-like units. Actually, raffinose is composed by a sucrose molecule with a galactose moiety bound to the O6 of the glucose subunit. We considered also the possibility that this feature arises from the furanoside moiety present in both the sugars: indeed, the lower flexibility of the penta-atomic with respect to the hexa-atomic ring [47] might constrain the protein structure in a new substate. However, the COB measured in MbCO embedded in fructose and ribose matrices did not show the above features, ruling out the interaction with furanose groups [26].

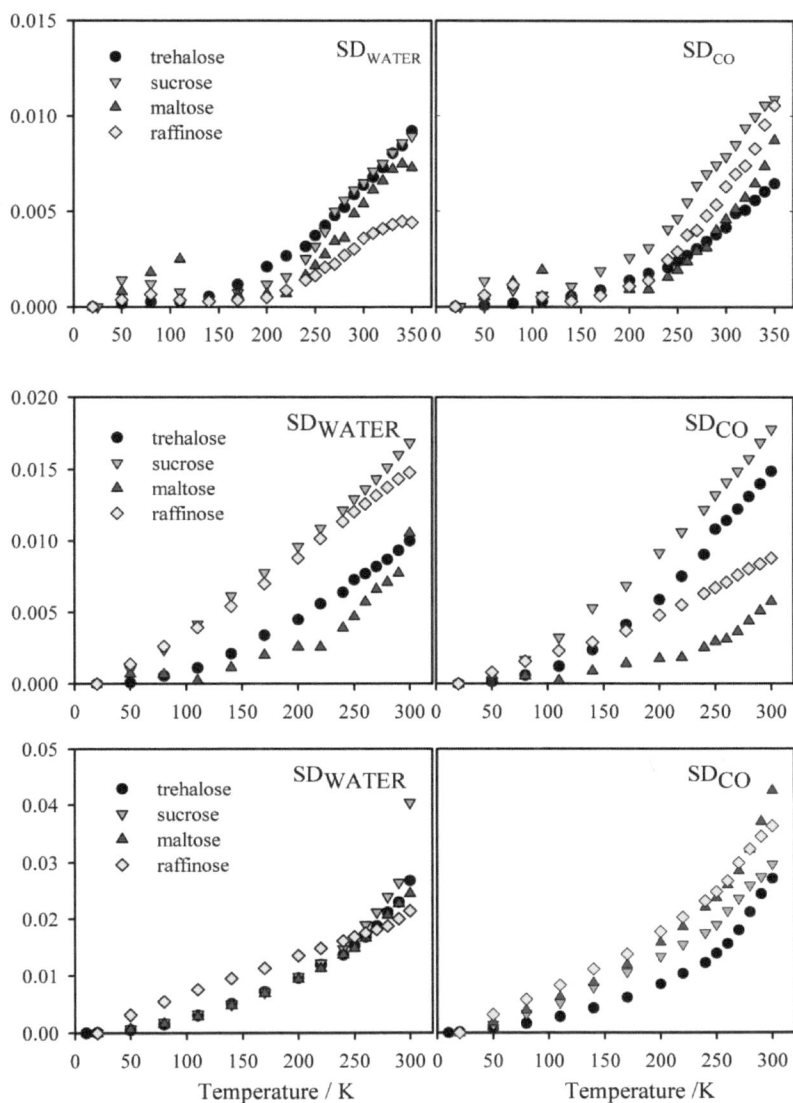

Figure 3: Spectra distance referred to the spectrum at 20K, calculated for the WAB (SD$_{WATER}$, left panels) and the COB (SD$_{CO}$, right panels). Data [21] refer to type 1 (very dry, upper panels), 2 (dry middle panels) and 3 (hydrated, lower panels) samples of trehalose (black circle), sucrose (grey down triangle), maltose (dark grey up triangle) and raffinose (light grey diamond). Note the difference of scale among the plots.

The thermal alterations of both the matrix and the embedded protein are shown in Fig. **3**, by using the SDs (see Eq. 1), calculated for the WAB (SD$_{WATER}$) and of the COB (SD$_{CO}$) and referred to the spectrum at 20 K. In very dry samples (Fig. **3**, upper panels) both SD$_{CO}$ and SD$_{WATER}$ have very low values up to ~200 K for all sugars. This indicates lack of structural changes in the protein as well as in the matrix and implies that, within our sensitivity, harmonic atomic motions dominate in both. Above 200 K a very small increase is evident for both SDs, stemming from interconversion among very low tiers substates. In dry samples (Fig. **3** middle panels), both SDs exhibit a roughly linear behavior, starting from very low temperature for sucrose, raffinose (~50K), and trehalose (~80K). The linear shape noticed in these saccharides complies with the presence of only low tier processes in the protein, coupled to rotational/vibrational fluctuations of water molecules. At variance from type 1 systems, such internal processes are already present at very low temperature, as reported also for dry lysozyme [48] and dry RNase powders [49]. The M2 sample has a rather different behaviour, exhibiting rather low values notwithstanding its large water content with respect to the other sugars: SD$_{WATER}$ for M1 and M2 samples overlap in the whole temperature range, while only slight differences are present between the two SD$_{CO}$ at high temperature. This can be explained on the basis of recent MD simulations results [51], which suggest that maltose-water solutions are inhomogeneous systems in which maltose molecules may cluster, likely due to the quite large dipole moment of this disaccharide [17]. It is conceivable that the water molecules are blocked at such clusters to the same extent as in a very dry system. The less constrained water molecules at the protein surface could instead be responsible for the slight differences between SD_{CO} [21].

At variance, the SD_{CO} plots for the same samples are rather linear up to 200 K and change slope above such temperature for all sugars, including raffinose. The low temperature behavior indicates that the hindrance of translational motions must involve also water at the protein-matrix interface. Furthermore, the different SD_{CO} plots are not overlapping over the whole temperature range, in line with the different protein-water-sugar structures observed by MD simulation in different sugar systems (see below and [28,50]). At higher temperature, after the collapse of the HB network, translational displacement of protein-bound water enables large-scale substates interconversion and protein structural relaxations, related to protein surface motions [20,29,30,41].

The raffinose behavior, with the slope change in SD_{CO} but not in SD_{WATER}, can be rationalized by considering that large-scale protein fluctuations are linked to translational motions at the interface, which may not extend to the whole matrix. Since the protein-bound water is only a fraction of the overall water content, its signal might be covered. Such effect is relevant only in raffinose due to the strong interactions this sugar has with water, which are expected to make rather different the behaviour of the two classes of water molecules.

MOLECULAR DYNAMICS

As reported above, FTIR data on MbCO embedded in the homologues disaccharide [26] pointed out that protein-matrix coupling depends in a subtle way on the detailed solvent composition. To rationalize at molecular level the effects of homologues disaccharides on the protein structure/dynamics and on the protein-solvent coupling, MD simulations were performed on MbCO embedded in sucrose or maltose-water box at 89% (sugar/[(sugar+water)] w/w, <2.3 water/sugar). Protein atomic fluctuations, protein-solvent at the protein interface, and solvent-solvent hydrogen bonding were analysed and compared with the one in trehalose [24,27]. A detailed comparison of atomic fluctuations along the chain (Fig. **4** upper panel) pointed out that, in sucrose and maltose, the residues with larger flexibility are located at the E, F, and H helices as well as at the EF and GH loops. Vitkup *et al.* [52] performed a MD simulation of MbCO in water in which water molecules were kept at 180 K, while the embedded protein was maintained at 300 K; this mimicked a "high viscosity solvent" uniformly surrounding the protein, whose MSFs resulted of very low amplitude and uniform over the whole chain. The MSFs relative to the simulation in trehalose, shown in Fig. **4**, upper panel, are most similar to the ones reported by Vitkup *et al.* in the case of "cold solvent". The similitude suggests, in line with FTIR results, that MbCO experiences stronger and more uniform constraint from the external matrix in trehalose than in sucrose or maltose; this in turn agrees with the higher bioprotective efficiency of trehalose. The atomic fluctuations of the protein embedded in saccharide-water matrices are also compared with the ones of a fully hydrated carboxy-myoglobin at 300 and 400 K, *i.e.* under conditions in which thermal denaturation takes place (see Fig. **4**, lower panel). As evident, some protein regions appear more involved in denaturation (E, F, and G helix residues, EF and GH loops). Fig. **4**, upper panel, also shows that motions in these parts of the protein are more hindered in trehalose than in sucrose or maltose. Results above-discussed on MSFs, along with the experimental results so far reported, lead to suggest that bioprotection should be discussed in the framework of protein-solvent coupling. Then, to understand the effects of sugar matrices on the structure and dynamics of the protein, it is necessary to determine the distribution of the mixed solvent at the macromolecule interface. In this respect, atomic-resolution molecular dynamics is an alternative, and complementary, method to experiments.

Water and sugar molecules were classified, in the configurations generated by MD, according to the distance r of the water oxygen or of the sugar hydroxyl group oxygen from the closest protein atom. We recall that each sugar molecule contributes eight OH groups. We calculated the ratio $N_{OW}/(N_{OW}+N_{OS})$ for each sugar, where N_{OW} and N_{OS} indicate, respectively, the number of water oxygen atoms and sugar hydroxyl groups oxygen atoms found at distance r from the protein surface, irrespective of their binding state. The second and third shells from the protein surface, which contain oxygen atoms whose distance from the protein is in the 1.0-3.0 Å interval (*i.e.*, atoms which could exchange hydrogen bonds with the protein sites), are richer in water oxygen atoms than in sugar oxygen atoms. The value of the ratio $N_{OW}/(N_{OW}+N_{OS})$ in the second and third shells is sizably higher than the asymptotic value of 0.22

in each sugar system. Accordingly, we suggest that, even in systems at 89% w/w, water preferentially surrounds the protein. This is in agreement with the model proposed by Timasheff [8] for bioprotection in dilute sugar solutions or preferential hydration model, *i.e.*, the solvent composition of the local protein domain is enriched in water relative to the average solvent composition in the outer shells (r > 4.0 Å).

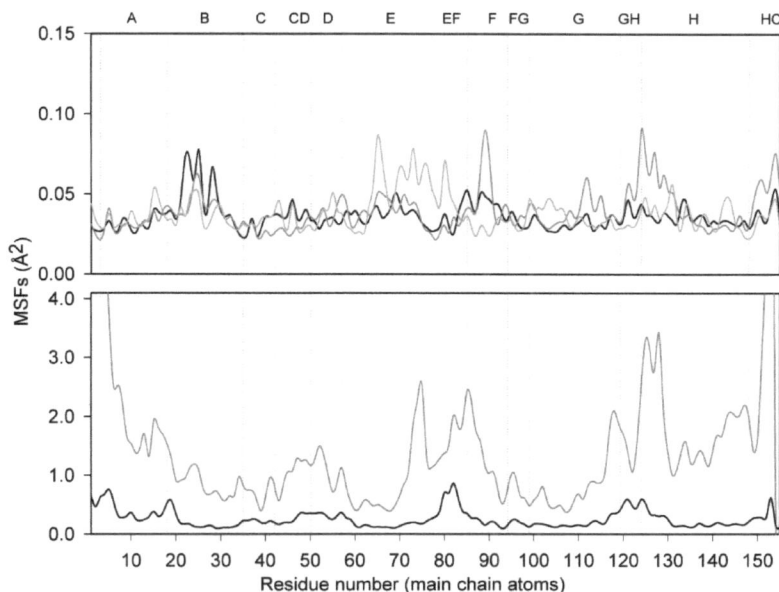

Figure 4: Atomic mean square fluctuations as a function of residue number, averaged over the protein main chain atoms (C, C α, N). Upper panel: MbCO in 89% w/w trehalose-water (black line); MbCO in 89% w/w sucrose-water (dark-grey line); MbCO in 89% w/w maltose-water (grey line). Lower panel: MbCO in water at 300 K (black line); MbCO in water at 400 K (dark-grey line). The labels on the top of the plot indicate the secondary structure elements of MbCO [24,27,28,50].

Trehalose, sucrose, and maltose molecules are bound to protein mainly through only one out of the eight OH groups available for binding (58%, 63%, 58% in trehalose, sucrose, and maltose, respectively). While the protein exchanges either single or multiple hydrogen bonds with water, the few sugar OH groups bound to the protein are in large majority linked through a single (donor) hydrogen bond. Results from simulations are in agreement with previous experimental data [53] which suggested that the interaction of trehalose or sucrose with the protein is restricted to sites assumed to loosely bind water molecules, which result more mobile. It was shown that the sugar-protein interaction does not affect the number of primary bound water molecules, which are likely to be tightly bound to the protein by multiple hydrogen bonds. Furthermore, along each simulated trajectory, the sugar molecules are found mainly bound to the carboxylate oxygen atoms of glutamic and aspartic residues as well as to the peptide oxygen atoms, in agreement with IR data on trehalose-protein systems [28]. Recent XAFS studies of cytochrome c in trehalose glasses at different content of residual water [54] pointed out a reduction of the mean square relative displacements of the central iron atom with respect the first coordination shell and a distortion of the porphirin group. The authors suggest that the matrix induces conformational changes in the protein, which in turn result in structural distortions of the heme group at the local atomic scale. Results from simulations show that a trehalose and a maltose molecule are bound to the heme propionate oxygen atoms in the <90% and 76% of the total number of configurations saved along the trajectories, respectively; this suggests that the heme could be involved in a direct interaction with the matrix. At variance, a sucrose molecule was found bound to the heme propionate oxygen atoms only in the 15% of the total number of saved configurations.

The protein-sugar-water systems simulated are mostly comparable to the humid (type 3) samples studied by FTIR experiments; in such systems, the matrix thermal structural rearrangements are almost overlapping in the three disaccharides, while differences are evident in the heme pocket thermal behavior. Such differences could be then ascribed to the structure of water at the interface. The fraction of protein-bound water molecules which is also bound to the sugar (or bridging water molecules), over the total number of protein-bound water, is higher in trehalose

(73%) than in sucrose (66%) or maltose (68%). This fraction of water molecules shared between protein and sugar could determine the strength of the constraints imposed by the matrix on the protein; then, this fraction could rationalize the larger local freedom for the protein internal motions in sucrose and maltose with respect to trehalose.

The fraction of water molecules bridging different sugar molecules, irrespective of their interaction with the protein, is 70%, 66%, and 63% for trehalose, maltose, and sucrose, respectively; this suggests that trehalose is more able to form structures in which water and sugar molecules cross-connect through the whole system [29], *i.e.* even far from the protein surface. The result is in line with simulative and experimental data on binary mixtures, indicating that the presence of trehalose sizably modifies the hydrogen bond network and the water dynamics by tightly binding water molecules [12,55]. In particular, a comparison with sucrose points out that, due to the more likely formation of internal hydrogen bonds, sucrose is more rigid and less hydrated than trehalose because of the lower number of sites available for interactions with water [44]; maltose, notwithstanding the similarity in hydration number, restrains the surrounding water molecules less than trehalose [56].

SMALL ANGLE X-RAY SCATTERING

The above reported results point out the presence of reciprocal protein-matrix effects from both the points of view of the local structure, at the molecular level, and of the dynamics of the protein and the matrix. However the reported data do not provide information on larger structural alteration of the samples and in particular, because of their spatial scale, they do not allow to mark the presence of micro- or mesoscopic inhomogeneities in this type of samples. We therefore performed SAXS measurements on binary and ternary trehalose and sucrose systems with the aim to detect the effects of the embedded MbCO on a larger spatial scale. SAXS was chosen since it is a most suitable technique to investigate micro-nano structures having electronic densities that differ from their surrounding.

Fig. **5**, left panel shows the SAXS patterns of low-water trehalose and sucrose glasses not containing protein, and of low-water myoglobin-trehalose and myoglobin-sucrose samples. Some feature of these systems can be directly extracted from the figure. A q^{-4} behavior, which is already evident at the lower bound of the measured q range, in all the samples indicates the presence of domains of, at least, micrometric size. This behaviour is present in a narrower range in sucrose samples, indicating the occurrence of very small inhomogeneities in this latter sugar, stemming from a possible formation of nanocrystals. This is further confirmed by the occurrence of a small peak at q ~0.15 Å^{-1} in the pattern of binary sucrose sample. This effect could be expected by considering that sucrose samples have a low residual water content, with respect to the other sugar, this promoting crystallisation, since sucrose crystallizes in anhydrous form [57].

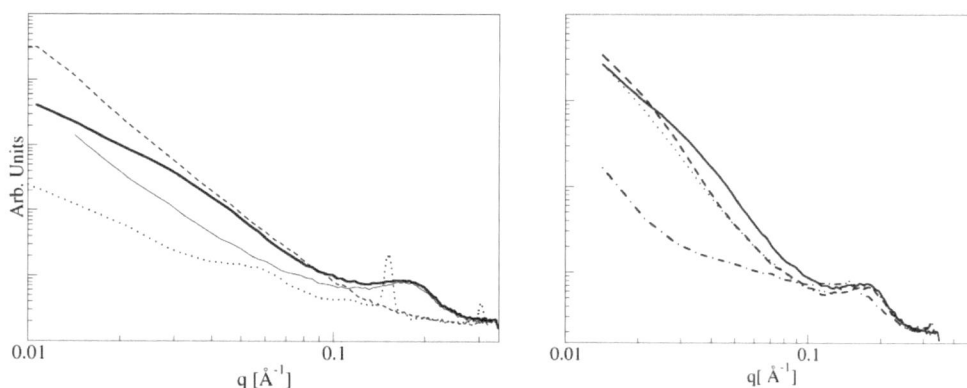

Figure 5: SAXS experimental patterns. The details of sample preparation and experimental condition are reported in [46]. Left panel: ternary trehalose-MbCO dry matrix (solid line), ternary sucrose-MbCO dry matrix (thin line), binary trehalose dry matrix (dashed line), binary sucrose dry matrix (dotted line). Right panel: ternary trehalose-MbCO matrix at various hydrations: the water content is increasing along the order solid line (dry sample, as in the left panel), dotted line, dashed line, dot-dashed line.

In the case of protein-containing samples a signal, originating from the protein, is evident in the 0.1-0.4 Å^{-1} q range. Myoglobin size resulted ~14 Å in both the sugar matrices, in agreement with the literature [58]. Because of the high protein concentration, a peak, arising from interference among Mb molecules occurs at $\approx 0.3 \text{Å}^{-1}$, corresponding to an average distance of ~31Å in both sugar, calculated with the Percus-Yevick function in decoupling approximation, with a spherical form factor for the protein [59].

In protein-containing trehalose samples only, it has been possible to detect a feature within the $0.01\text{-}0.1 \text{ Å}^{-1}$ range, corresponding to a broad distribution of structures. A fitting with the Debye function, assuming additivity, gave a broad distribution of domains of average size ~150 Å. Taking into account the average dimensions and the concentration of the three components of the system, the overwhelming volume fraction of the sample must be constituted by myoglobin molecules. It follows that protein-rich regions are the background against which a signal, from protein-poor domains, can be observed [46].

Upon hydration (see Fig. **5**, right panel) the protein background shows initially little difference with respect to dry samples, as pointed out from the unaltered protein–protein distance and protein size. This could be rationalized by considering that the incoming water must preferentially concentrate at the more hydrophilic "sugar rich/protein poor" domains. Further hydration makes the domains signal to disappear and the protein signal to broaden and shift. We attribute such behaviour to a change in the structure of the sample: the larger water availability makes to increase the hydration of the protein and leads to the merging of the domains with the background. The new computed distance among protein molecules reach ~45Å, which is compatible with the presence of at least one hydration layer surrounding the protein [46].

The model exploited from SAXS data depicts, for low water binary samples, a rather homogeneous system with extended HB networks in trehalose, and a less homogeneous system containing nanocrystals in sucrose. These nanocrystals form because of the increase in the fraction of intramolecular HB in sucrose molecules, which, reducing the connectivity of the network, promotes formation of crystalline anhydrous sucrose. At variance, the model for ternary systems points out, in trehalose, the presence of protein-poor domains embedded in a protein-rich background. In the latter the protein is locked into its environment through a hydrogen bond network involving trehalose and water, whereas the protein-poor domains are mainly constituted by regions of trehalose and water connected by extensive HB networks. The occurrence of domains is likely due to the system stabilisation caused by the partial seclusion of the strong HB-maker trehalose from the weak HB-maker (chaotropic) protein. In ternary sucrose systems, at variance, the stronger competition of internal hydrogen bonding leads to the absence of seclusion and thus to the relative stabilization of non-compartmentalised systems. The presence of a major perturbation of the system in trehalose ternary matrices well agrees with the FTIR results (see above, Fig. **1** panel *a*), which show that trehalose, at variance with sucrose, strongly alter the WAB shape upon protein addition, hence it can strongly affect the matrix structure and behaviour [46]. In our opinion, the occurrence of inhomogeneities should therefore have origin in the properties of the HB networks established in the systems. Unduly importance should not be credited to the specific nature of the saccharide, the main properties relying in the HB networks it form, which could also be modulated by other components of the systems as well as other drying procedures.

CONCLUSIONS

The results obtained with the presented techniques allow to suggest a description of the solid amorphous sugar matrices, of their mechanism of preservation and of the peculiarity of trehalose. FTIR spectra, MD simulations and, on a larger space scale, SAXS measurements, suggests the presence of extended water-saccharide HB networks, whose rigidity is modulated by water content. The thermal behaviour of the embedded protein and of the matrix result strongly coupled and both the components concur in defining the whole system properties. The comparison among various sugars matrices points out that, despite the structural similarity, saccharides behaves differently in their interaction with the other components; in this context, the better biopreserving properties of trehalose are explained in terms of stronger interactions both with water and the protein leading to a stronger protein-matrix coupling.

REFERENCES

[1]   Crowe LM. Lesson from nature: a role of sugars in anhydrobiosis. Comp Biochem Physiol A 2002; 132: 505–13.

[2]   Crowe JH. Trehalose as a "chemical chaperone": fact and fantasy. Adv Exp Med Biol 2007; 594: 143–58.

[3]   Carpenter JF, Crowe JH. An infrared spectroscopic study of the interaction of carbohydrates with dried proteins.

Biochemistry 1989; 28: 3916–22.

[4]   Sum AK, Faller R, de Pablo JJ. Molecular simulation study of phospholipid bilayers and insights of the interactions with disaccharides. Biophys J 2003; 85: 2830–44.

[5]   Villarreal MA, Díaz SB, Disalvo EA, Montich GG. Molecular dynamics simulation study of the interaction of trehalose with lipid membranes. Langmuir 2004; 20: 7844–51.

[6]   Pereira C, Hunenberger PH. Interaction of the sugars trehalose, maltose and glucose with a phospholipid bilayer: a comparative molecular dynamics study. J Phys Chem B 2006; 110: 15572–81.

[7]   Belton PS, Gil AM. IR and Raman spectroscopic studies of the interaction of trehalose with hen egg white lisozyme. Biopolymers 1994; 34: 957–61.

[8]   Timasheff SN. Protein hydration, thermodynamic binding, and preferential Hydration. Biochemistry 2002; 41: 13473–82.

[9]   Sampedro JG, Uribe S, Trehalose-enzyme interactions result in structure stabilization and activity inhibition. Mol Cell Biochem 2004; 256: 319–27.

[10]   Green J, Angell CA. Phase relations and vitrification in saccharide-water solutions and the trehalose anomaly. J Phys Chem 1989; 93: 2880–82.

[11]   Branca C, Magazù S, Maisano G, Migliardo P. Anomalous cryoprotective effectiveness of trehalose: Raman scattering evidences. J Chem Phys 1999; 111: 281–8.

[12]   Magazù S, Maisano G, Migliardo F, Mondelli C. Mean-square displacement relationship in bioprotectant systems by elastic neutron scattering. Biophys J 2004; 86: 3241–9.

[13]   Hédoux A, Willart JF, Ionov R, Affouard F, Guinet Y, Paccou L, Lerbret A, Descamps M. Analysis of sugar bioprotective mechanisms on the thermal denaturation of lysozyme from Raman scattering and differential scanning calorimetry investigations. J Phys Chem B 2006; *110:* 22886–93.

[14]   Sussich F, Skopec C, Brady J, Cesàro A. Reversible dehydration of trehalose and anhydrobiosis: from solution state to an exotic crystal? Carbohydr Res 2001; 334: 165–76.

[15]   Allison SD, Chang B, Randolph TW, Carpenter JF. Hydrogen bonding between sugar and protein is responsible for inhibition of dehydration-induced protein unfolding. Arch Biochem Biophys 1999, 365: 289–98.

[16]   Wolkers WF, Oldenhof H, Alberda M, Hoekstra FA. A Fourier transform infrared microspectroscopy study of sugar glasses: application to anhydrobiotic higher plant cells. Biochim. Biophys. Acta-Gen Subj 1998; 1379: 83–96.

[17]   Gaffney SH, Haslam E., Lilley TH, Ward TRJ. Homotactic and heterotactic interactions in aqueous-solutions containing some saccharides experimental results and an empirical relationship between saccharide solvation and solute interactions. J Chem Soc Faraday Trans 1998; 84: 2545–52.

[18]   Hagen SJ, Hofrichter J, Eaton WA. Protein reaction kinetics in a room temperature glass. Science 1995; 269: 959–62.

[19]   Gottfried DS, Peterson ES, Sheikh AG, Wang J, Yang M, Friedman JM. Evidence for damped hemoglobin dynamics in a room temperature glass. J Phys Chem 1996; 100: 12034–42.

[20]   Cordone L, Cottone G, Giuffrida S, Palazzo G, Venturoli G, Viappiani C. Internal dynamics and protein-Matrix coupling in trehalose coated proteins. BBA-Proteins Proteomics 2005; 1749: 252–81 (and references therein).

[21]   Giuffrida S, Cottone G, Cordone L. Role of solvent on protein-matrix coupling in MbCO embedded in water-saccharide systems: a Fourier Transform Infrared spectroscopy study. Biophys J 2006; 91: 968–80.

[22]   Cordone L, Cottone G, Giuffrida S. Role of residual water hydrogen bonding in sugar/water/biomolecule systems: a possible explanation for trehalose peculiarity. J Phys-Condens Matter 2007; 19: 205110 (and references therein).

[23]   Cordone L, Cottone G, Giuffrida S, Librizzi F. Thermal evolution of the CO stretching band in carboxy-myoglobin in the light of neutron scattering and molecular dynamics simulations. Chem Phys 2008; 345: 275–82.

[24]   Cottone G, Cordone L, Ciccotti G. Molecular Dynamics simulation of carboxy-myoglobin embedded in a trehalose-water matrix. Biophys J 2001; 80: 931–8.

[25]   Francia F, Palazzo G, Mallardi A, Cordone L, Venturoli G. Probing light-induced conformational transitions in bacterial photosynthetic reaction centers embedded in trehalose amorphous matrices. Biochim Biophys Acta-Bioenerg 2004; 1658: 50–7.

[26]   Librizzi F, Viappiani C, Abbruzzetti S, Cordone L. Residual water modulates the dynamics of the protein and of the external matrix in trehalose coated MbCO: an infrared and flash-photolysis study. J Chem Phys 2002; 116: 1193–200.

[27]   Cottone G, Ciccotti G, Cordone L. Protein-trehalose-water structures in trehalose coated carboxy-myoglobin. J Chem Phys 2002, 117 9862–6.

[28]   Cottone G. A comparative study of carboxy myoglobin in saccharide–water systems by Molecular Dynamics simulation. J Phys Chem B 2007; 111: 3563–9.

[29]   Giuffrida S, Cottone G, Librizzi F, Cordone L. Coupling between the thermal evolution of the heme pocket and the external matrix structure in trehalose coated carboxymyoglobin. J Phys Chem B 2003; 107: 13211–7.

[30]   Giuffrida S, Cottone G, Cordone L. Structure-dynamics coupling between protein and external matrix in sucrose-coated and in trehalose-coated MbCO: an FTIR study. J Phys Chem B 2004; 108: 15415–21.

[31]   Frauenfelder H, Parak F, Young RD, Conformational substates in proteins, Ann Rev Biophys Chem 1988; 17: 451–79.

[32]   Vojtěchovský J, Chu K, Berendzen J, Sweet RM, Schlichting I. Crystal structures of myoglobin-ligand complexes at near-atomic resolution. Biophys J 1999; 77: 2153–74.

[33]   Beece D, Eisenstein L, Frauenfelder H, Good D, Marden MC, Reinisch L, Reynolds AH, Sorensen LB, Yue KT. Solvent viscosity and protein dynamics. Biochemistry 1980; 19: 5147–57.

[34]   Eisenberg D, Kauzmann W, The structure and properties of water, Oxford University Press, London, 1969.

[35]   Giuffrida S, Cottone G, Vitrano E, Cordone L. A FTIR study on low hydration saccharide amorphous matrices: thermal behaviour of the Water Association Band. J Non-Cryst Solids, submitted.

[36]   Francia F, Dezi M, Mallardi A, Palazzo G, Cordone L, Venturoli G. Protein-matrix coupling/uncoupling in "dry" systems of photosynthetic Reaction Center embedded in trehalose/sucrose: the origin of trehalose peculiarity. J Am Chem Soc 2008; 130: 10240–6.

[37]   Samuni U, Dantsker D, Roche CJ, Friedman JM. Ligand recombination and a hierarchy of solvent slaved dynamics: the origin of kinetic phases in hemeproteins. GENE 2007; 398: 234–48, and references therein.

[38]   Massari AM, Finkelstein IJ, Fayer MD. Dynamics of proteins encapsulated in silica sol-gel glasses studied with IR vibrational echo spectroscopy. J Am Chem Soc 2006; 128: 3990-7.

[39]   Doster W, Bacheitner A, Dunau R, Hiebl M, Luscher E. Thermal properties of water in myoglobin crystals and solutions at subzero temperatures. Biophys J 1986; 50: 213-9.

[40]   Tarek M, Tobias DJ. Role of Protein-Water Hydrogen Bond Dynamics in the Protein Dynamical Transition. Phys Rev Lett 2002; 88: 138101-6.

[41]   Rector D, Jiang J, Berg MA, Fayer MD. Effects of solvent viscosity on protein dynamics: infrared vibrational echo experiments and theory. J Phys Chem B 2001; 105: 1081-92.

[42]   Heyden M, Bründermann E, Heugen U, Niehues G, Leitner DM, Havenith M. Long-range influence of carbohydrates on the solvation dynamics of waters - answers from terahertz absorption measurements and molecular modeling simulations. J Am Chem Soc 2008; 130: 5773–9.

[43]   Bellavia G, Cottone G, Giuffrida S, Cupane A, Cordone L. Thermal Denaturation of Myoglobin in Water-Disaccharide Matrixes: Relation with the Glass Transition of the System. J Phys Chem B 2009; 113: 11543-9.

[44]   Ekdawi-Sever NC, Conrad PB, de Pablo JJ. Molecular simulation of sucrose solutions near the glass transition temperature. J Phys Chem A 2001; 105: 734–42.

[45]   Berman HM. The crystal structure of a trisaccharide, raffinose pentahydrate. Acta Cryst 1970; B26: 290–9.

[46]   Longo A, Giuffrida S, Cottone G, Cordone L. Myoglobin embedded in saccharide amorphous matrices: water-dependent domains evidenced by Small Angle X-Ray Scattering. Phys Chem Chem Phys 2010; doi:10.1039/b926977k

[47]   Rao VSR, Qasba PK, Balaji PV, Chandrasekaran R. Conformation of Carbohydrates. Harwood Academic Publishers, 1998, Newark.

[48]   Caliskan G, Briber RM, Thirumalai D, Garcia-Sakai V, Woodson SA, Sokolov APJ. Dynamic transition in tRNA is solvent induced. J Am Chem Soc 2006; 128: 32–3.

[49]   Curtis JE, Tarek M, Tobias DJ. Methyl group dynamics as a probe of the protein dynamical transition. J Am Chem Soc 2004; 126: 15928–9.

[50]   Cottone G, Giuffrida S, Ciccotti G, Cordone L. Molecular dynamics simulation of sucrose- and trehalose-coated carboxy-myoglobin, Proteins 2005; 59: 291–302.

[51]   Lerbret A, Bordat P, Affouard F, Descamps M, Migliardo F. How homogeneous are the trehalose, maltose, and sucrose water solutions? An insight from molecular dynamics simulations. J Phys Chem 2005;109: 11046–57.

[52]   Vitkup D, Ringe D, Petsko GA, Karplus M. Solvent mobility and the protein 'glass' transition. Nat Struct Biol 2000; 7: 34-8.

[53]   Lopez-Diez EC, Bone S. An investigation of the water-binding properties of protein + sugar systems. Phys Med Biol 2000; 45: 3577- 88.

[54]   Giachini L, Francia F, Cordone L, Boscherini F, Venturoli G. Cytochrome c in a Dry Trehalose Matrix: Structural and Dynamical Effects Probed by X-Ray Absorption Spectroscopy. Biophys J 2007; 92: 1350-60.

[55]   Conrad PB, de Pablo JJ. Computer Simulation of the Cryoprotectant Disaccharide α,α-Trehalose in Aqueous Solution. J Phys Chem A 1999; 103: 4049- 55.

[56]   Sakurai M, Murata M, Inoue Y, Hino A, Kobayashi S. Computer Simulation of the Cryoprotectant Disaccharide α,α-Trehalose in Aqueous Solution. Bull Chem Soc Jpn 1997; 70: 847-58.

[57]   Brown GM, Levy HA. Further refinement of the structure of sucrose based on neutron-diffraction data. Acta Cryst 1973 ; B29: 790-7.

[58] Ibel K, Stuhrmann HB. Comparison of neutron and X-ray scattering of dilute myoglobin solution. J Mol Biol 1975; 93: 255-66.

[59] Pedersen JS, Analysis of small-angle scattering data from colloids and polymer solutions: modeling and least-squares fitting, Adv Colloid Interface Sci 1997; 70: 171-210.

CHAPTER 6

Sugar – Lipid Interactions: Structural and Dynamic Effects

Antonio Deriu[*], Maria Teresa Di Bari and Yuri Gerelli

Dipartimento di Fisica, Università degli Studi di Parma and CNISM, Viale G.P. Usberti 7/a, Parma, 43100, Italy

Abstract: Sugar-lipid complexes are nowadays extensively applied in biology, medicine and food science. In particular, sugars play an important role in maintaining cell viability in stressful environmental conditions. A detailed understanding of lipid-saccharide-solvent interactions can be achieved by a combined use of advanced microscopic structural and dynamical investigation techniques. In this review the effect of saccharide content on the gel to liquid–crystalline phase transition and on the multilayer structure of lipid membranes, as well as on the aggregation properties of liposomes in colloidal systems, is discussed.

INTRODUCTION

In the last years, many studies have been devoted to the investigation of the interactions between sugars and other biomolecules owing to the bio-protective role played by some saccharides [1]. This behaviour has been explained in terms of the glass forming properties of water-saccharide systems [2] and of their *water substitution* capability [3]. Sugar–water interactions are therefore a key item and they have been studied in detail using a variety of spectroscopic techniques: infrared spectroscopy [4], elastic- [5] and quasielastic neutron scattering [6], Rayleigh scattering of Mössbauer radiation [7], as well as by molecular dynamics simulations [8]. All these studies indicate that the structure and the dynamics of water are affected by the interaction with the saccharides.

Among sugar-biomolecule complexes, sugar–protein and sugar–lipid systems have been studied since the late eighties [9, 10]. In the case of sugar–protein interactions the most relevant phenomenon is a marked increase in protein thermal and conformational stability when sugars are present in protein-water solutions [11]. Sugar-lipid complexes are nowadays extensively applied in biology [12], medicine and food science [11]. In particular, sugars play an important role in maintaining cell viability under dry conditions, and thus enable organisms to survive in the dry state (anhydrobiosis) and in stressful environmental conditions (typical examples are tardigrades, nematodes, and brine shrimps) [13-16].

Saccharide-stabilized bilayers have been investigated using Fourier transform infrared spectroscopy, Raman spectroscopy, nuclear magnetic resonance and differential scanning calorimetry [17–19]. Sugar-lipid interactions decrease the gel to liquid–crystalline phase transition temperature observed in dry bilayers keeping dry membranes in a physical state similar to that of hydrated ones [14,19]. The protective role of sugars is also due to their ability to form a rigid glassy matrix at ambient temperature avoiding membrane fusion [17]. The effect of H–bond interactions between sugars and lipids can be enhanced using charged saccharides together with oppositely charged lipids. Addition of polyelectrolytes results in modifications of the surface charge of lipid membranes, as it results from ζ–potential measurements [20], and consequently on some surface properties as for instance bioadhesion.

Addition of charged saccharides modifies not only the dynamical properties of lipid assemblies but also their structure [21]. The structural features strongly depend upon the polysaccharide content; they have been investigated using dynamic light scattering (DLS), small angle neutron (SANS) and X–ray (SAXS) scattering. Optic, transmission electron microscopies have also been used to obtain direct information on the morphology of such complexes. When the charge of the adsorbed polyelectrolytes neutralizes the original charge of the particle surface, *i.e.* close to the isoelectric point of the solution, large clusters are observed. A further increase of the polyelectrolyte–particle ratio progressively reduces the size of the aggregates (re–entrant condensation [22– 24]).

SUGAR–LIPID INTERACTIONS

Gel to Liquid–Crystalline Phase Transition

Saccharide addition decreases significantly the gel to liquid–crystalline phase transition temperature, T_m, observed in phospholipid vesicles [14, 19]. As a consequence, phase transitions are avoided during dehydration and rehydration at

---

*Address correspondence to Antonio Deriu: Dipartimento di Fisica, Università degli Studi di Parma and CNISM , Viale G.P. Usberti 7/a, Parma, 43100, Italy, Antonio.Deriu@fis.unipr.it

Salvatore Magazù and Federica Migliardo (Eds)

temperatures close to ambient, preventing thus the loss of internal solutes. The decrease of T_m can be easily observed by DSC. Fig. **1a** shows heating thermograms of dry egg phosphatydlcholine (EPC) liposomes in presence of different amounts of sucrose. The shift of the endothermic peak towards lower temperatures as the sugar/lipid ratio increases is clearly visible.

NMR and FTIR investigations provide more insight on the sugar induced stabilization of lipid bilayers. Detailed ^{2}H and ^{13}C solid state NMR investigations have been carried out on bilayers made up from the model phospholipid DPPC (1,2-dipalmitoyl-sn-phosphatidylcholine) in presence a matrix of sugar (trehalose) with an overall 2:1 trehalose/DPPC concentration. [17]. The NMR spectra of trehalose-stabilized bilayers showed significant differences with respect to those of hydrated bilayers, while revealing a number of interesting similarities. The interaction of trehalose with the hydrophilic headgroups of the phospholipids inhibits axial diffusion of the lipid molecules in the bilayer. In contrast, the hydrocarbon chains exhibit extensive trans-gauche isomerization. This chain disorder can be explained if one postulates a bilayer structure where the trehalose is intercalated between lipid molecules, thus acting as a spacer to expand the bilayer.

FTIR spectroscopy can probe the dynamics of the different parts of the lipid component; through the analysis of some characteristic frequencies due to, for instance, phosphate asymmetric stretching, carbonyl stretching, and asymmetric stretching of the choline C–N bond. As an example, Fig. **1b** shows the wavenumber of the symmetric CH$_2$ stretching band (v CH$_2$s) as a function of temperature for dry EPC at different sucrose contents: the obtained curves confirm the depression of T_m induced by addition of sucrose. The shift in T_m between pure EPC liposomes and the sample with the highest content of sucrose is about 60 °C. The FTIR data show that sucrose interacts, thorough H–bonding, with the polar P=O and C=O groups of the lipid and also with the methyl groups of the choline moiety; these interactions are qualitatively very similar to those between water and lipids. Quantitatively, the strength of the interaction is similar only at the level of the choline groups, while it is weaker for the polar headgroups. This scenario confirms the ability of sugar to replace hydration water on the lipid surface. Moreover the formation of a rigid and stable glassy matrix prevents vesicle membrane fusion at sufficiently high sugar–lipid ratios if the sugar is present both inside and outside the vesicles.

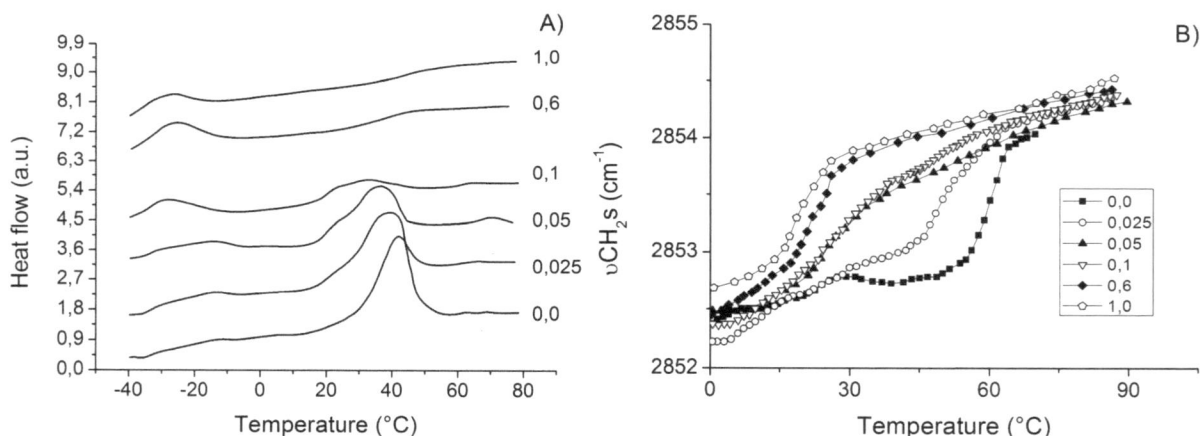

Figure 1: a) DSC heating thermograms of dry EPC liposomes in presence of different amounts of sucrose; b) lipid melting curves of dry EPC liposomes with different amounts of sucrose as determined by FTIR spectroscopy. The wavenumber of the symmetric CH$_2$ stretching band (vCH$_2$s) is plotted vs. temperature. In both panels the saccharide-to-lipid mass ratios are indicated for the different curves. Readapted with permission from [19].

ELECTROSTATIC EFFECTS

The capability of saccharides to stabilize lipid structures can be exploited in biomedical and pharmaceutical applications as the development of new drug delivery systems. A typical example is that of chitosan–lecithin nanoparticles [21]. In this case the saccharide–lipid interaction is not only due to H–bonding; it originates from electrostatic effects between the positively charged polysaccharide (chitosan) and the negatively charged lipids (lecithin). Fig. **2a** shows the SAXS pattern for pure lipid particles; in this case the signal is predominantly due to unilamellar vesicles (*i.e.* with a hollow core surrounded by a single lipid bilayer) whose form factor is shown as a dashed line. The presence of a small population of multilamellar vesicles gives rise to a broad additional contribution centred at $Q \approx 0.11$ Å^{-1}. Addition of chitosan increases the number of multilamellar vesicles. As a consequence the peak at $Q \approx 0.11$ Å^{-1} is more evident (Fig. **2b**). Its width becomes narrower with increasing chitosan content indicating the presence of higher order multilayer repetitions, although in a low number fraction.

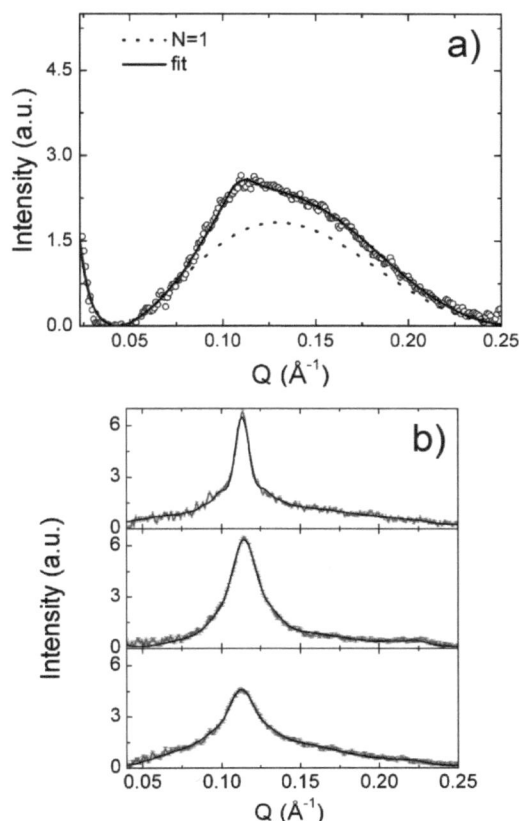

Figure 2: a) SAXS profile of pure lecithin vesicles. The total fit (solid line) is shown together with the form factor of the unilamellar vesicles (dashed line). b) SAXS curves for chitosan–lecithin nanoparticles at different chitosan/lecithin ratios (1/80, 1/20, 1/5 (w/w) top to bottom). The peak due to the multilamellar repetition is clearly visible. Readapted with permission from [21].

It is worth remarking that multilamellarity affects also the kinetics of drug release since a higher number of confining layers has to be crossed (or dissolved). In this view a fine tuning of the multilayer structure controlled by the amount of sugar present can play a key role in optimizing nanoparticle properties for drug delivery applications.

CLUSTERING EFFECTS

Charged colloidal particles are able to self–assemble, when mixed in an aqueous solvent with oppositely charged polyelectrolytes, forming long–lived finite–size mesoscopic aggregates. Among colloidal preparations it is worth mentioning liposomal ones; in fact long–lived finite size lipid vesicles 'glued together' by oppositely charged polyelectrolytes show an interesting potential for developing a new class of multi-compartment vectors for the simultaneous delivery of different pharmacologically active molecules. The physical context in which the sugar–lipid complexation mechanism can be placed is that of macroion–polyion interactions [22]. Particular attention has been devoted to complexes formed by lipid vesicles incubated with polyions, adhering to their surfaces and giving rise to ordered patterns or coordinating different vesicles. A rich literature exists on complex structures consisting of amphiphile aggregates interacting *via* electrostatic forces with linear polyions, polypeptides, or DNA, including lamellar phases or dispersed particles [25,26]. A peculiar feature of the phase behaviour of such systems is charge inversion, corresponding to oppositely charge polyelectrolytes complexing beyond charge neutralization [22-24]. Amphiphile/polyion coordination reaches its maximum in the region of charge inversion where large clusters are formed. This behaviour can be described in terms of a re–entrant condensation effect: extra polyions can be attracted to the surface of the macroion despite the fact that this surface is already neutralized by the previously adsorbed polyions. This entropically driven process is due to an induced attraction between polyions [27]. The re-entrant condensation effect has been evidenced in uni-lamellar liposomes built up with a synthetic lipid, DOTAP (1,2-dioleoyl-3-trimethyl ammonium propane). Their aggregation is induced by adding to the suspension increasing amounts of an anionic synthetic polyelectrolyte, sodium polyacrylate (NaPA) [23]. Fig. **3** shows the average hydrodynamic radius and the surface potential of the lipid–polyion particles at different lipid–polyion ratios. Near the isoelectric point (ξ=1.1) there is an increase of the average particle size and of the inversion of the surface charge.

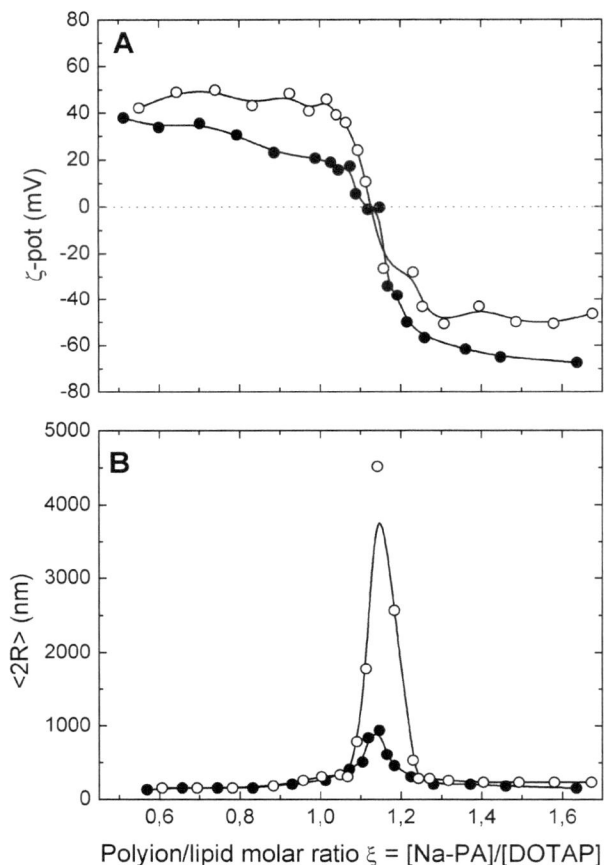

Figure 3: Re–entrant condensation (panel A, aggregate diameter) and charge inversion (panel B, ζ potential) of cationic DOTAP liposomes in the presence of the anionic NaPA polyelectrolyte. Data are shown as a function of the polyion–lipid molar charge ratio, ξ, for two different temperatures, 80°C (empty symbols) and 5°C (full symbols). Readapted with permission from [23].

CONCLUSIONS

Addition of saccharides to lipid layer systems significantly modifies the structure and morphology of lipid assemblies as well their dynamics at the microscopic level. These changes may have a relevant effect on the stability and permeability of lipid membranes. A detailed understanding of lipid-saccharide-solvent interactions can be achieved by a combined use of advanced microscopic structural and dynamical investigation techniques. Sugars interact mostly with the polar headgroups of the lipid molecules, but also with the methyl groups of the hydrocarbon chains. In this way they partly intercalate between lipid molecules acting as spacers to expand the lipid bilayers. Saccharide-stabilized lipid multilayers may play an important role in maintaining membrane structure and cell viability in extreme environmental conditions and in particular in the dehydrated state. In lipid based vesicles a fine tuning of the multilayer structure can be obtained by varying the amount of sugar present within the lipid bilayers. In this way the release properties of lipid-based nanoparticles can be optimized for medical and pharmacological applications. Using charged saccharides in colloidal preparations clustering and aggregation effects can be controlled acting on the surface charge of the polyelectrolyte-decorated particles. This interesting phenomenology has a high potential for biotechnological applications, for instance in developing multi-compartment vectors for the simultaneous intra-cellular delivery of different pharmacologically active substances.

REFERENCES

[1]   Jain NK, Roy I. Effect of trehalose on protein structure. Prot Sci 2008; 18: 24-36.

[2]   Green JL, Angell CA. Phase-relations and vitrification in saccharide-water solutions and the trehalose anomaly. J Phys Chem B 1989; 93: 2880–2.

[3]   Donnamaria MC, Howard EI, Grigera JR. Interaction of water with α-α trehalose in solution: Molecular dynamics simulation approach. J Chem Soc Faraday Trans 1994; 90: 2731–5.

[4]   Giangiacomo R. Study of water-sugar interactions at increasing sugar concentration by NIR spectroscopy. Food Chem 2006; 96: 371–9.

[5]   DiBari MT, Deriu A, Albanese G, Cavatorta F. Dynamics of hydrated starch saccharides. Chem Phys 2003; 292: 333–9.

[6]   Magazù S, Migliardo F, Telling MTF. α,α-Trehalose/Water Solutions. VIII. Study of Diffusive Dynamics of Water by High-Resolution Quasi Elastic Neutron Scattering. J Phys Chem B 2006; 110: 1020–5.

[7]   Deriu A, Cavatorta F, Albanese G. Rayleigh Scattering of Mössbauer Radiation in Hydrated Amylose. Hyperfine Interactions 2002; 141–142: 261–5.

[8]   Lerbret A, Bordat P, Affouard F, Hédoux A, Guinet Y, Descamps M. How Do Trehalose, Maltose, and Sucrose Influence Some Structural and Dynamical Properties of Lysozyme? Insight from Molecular Dynamics Simulations. J Phys Chem B 2007; 111: 9410–20.

[9]   Quiocho FA. Carbohydrate-binding proteins: tertiary structures and protein-sugar interactions. Ann Rev Biochem 1986; 55: 287–315.

[10]   Colaço C, Sen S, Thangavelu M, Pinder S, Roser B. Extraordinary stability of enzymes dried in trehalose: simplified molecular biology. Nature 1992; 10: 1007–11.

[11]   Semenova MG, Antipova AS, Belyakova LE. Food protein interactions in sugar solutions. Current Opinion in Colloid & Interface Science 2002; 7: 438–44.

[12]   Caffrey M, Fonseca V, Leopold AC. Lipid-sugar interactions. Plant Physiol 1988; 86: 754–8.

[13]   Crowe JH, Crowe LM, Carpenter JF, Wistrom CA. Stabilization of dry phospholipid bilayers and proteins by sugars. Biochem J 1987; 242: 1–10.

[14]   Crowe JH, Hoekstra FA, Crowe LM. Anhydrobiosis. Ann Rev Physiol 1992; 54: 579–99.

[15]   Oliver AE, Crowe LM, Crowe JH. Methods for dehydrationtolerance: depression of the phase transition temperature in dry membranes and carbohydrate vitrification. Seed Sci Res 1998; 8: 211–21.

[16]   Oliver AE, Hincha DK, Crowe JH. Looking beyond sugars: the role of amphiphilic solutes in preventing adventitious reactions in anhydrobiotes at low water contents. Comp Biochem Physiol 2002; 131A: 515–25.

[17]   Lee CWB, Das Gupta SK, Mattai J, Shipley GG, Abdel–Mageed OH, Makriyannis A, Griffin RG. Characterization of the L lambda phase in trehalose-stabilized dry membranes by solid-state NMR and X-ray diffraction. Biochemistry 1989; 28: 5000–9.

[18]   Strauss G, Hauser H. Stabilization of lipid bilayer vesicles by sucrose during freezing. PNAS 1986; 83: 2422–6.

[19]   Cacela C, Hincha DK. Low amounts of sucrose are sufficient to depress the phase transition temperature of dry phosphatidylcholine, but not for lyoprotection of liposomes. Biophys J 2006; 90: 2831–42.

[20]   Hunter RJ. Zeta Potential in Colloid Science: Principles and Applications, Academic Press: New York, 1981.

[21]   Gerelli Y, Barbieri S, DiBari MT, Deriu A, Cantù L, Brocca P, Sonvico F, Colombo P, May R, Motta S. Structure of self-organized multilayer nanoparticles for drug delivery. Langmuir 2008; 24: 11378–84.

[22]   Bordi F, Cametti C, Diociaiuti M, Gaudino D, Gili T, Sennato S. Complexation of anionic polyelectrolytes with cationic liposomes: evidence of reentrant condensation and lipoplex formation. Langmuir 2004; 20: 5214–22.

[23]   Sennato S, Truzzolillo D, Bordi F, Cametti C. Effect of temperature on the reentrant condensation in polyelectrolyte-liposome complexation. Langmuir 2008; 24: 12181–8.

[24]   Bordi F, Sennato S, Truzzolillo D. Polyelectrolyte-induced aggregation of liposomes: a new cluster phase with interesting applications. J Phys: Condens Matter 2009; 21: 203102-6.

[25]   Nguyen TT, Grosberg AY, Shklovskii BI. Charged surface in salty water with multivalent ions: Giant inversion of charge. Phys Rev Lett 2000; 85: 1568–71.

[26]　Dobrynin AV, Deshkovski A, Rubinstein M. Adsorption of Polyelectrolytes at an Oppositely Charged Surface. Phys Rev Lett 2000; 84: 3101–4.

[27]　Grosberg AY, Nguyen TT, Shklovskii BI. Colloquium: the physics of charge inversion in chemical and biological systems. Rev Mod Phys 2002; 74: 329–45.

SECTION IV

Simulation and Complementary Spectroscopic Techniques

CHAPTER 7

Studies of Biomacromolecules with Neutron Vibrational Spectroscopy

Stewart F. Parker[*]

ISIS Facility, STFC Rutherford Appleton Laboratory, Chilton, Didcot, Oxfordshire, OX11 0QX, United Kingdom

Abstract: The advantages of vibrational spectroscopy by the use of inelastic neutron scattering (INS) spectroscopy are illustrated for biomacromolecules and their interaction with water. The complementarity with other vibrational spectroscopic techniques is demonstrated and the synergy with calculations is stressed.

INTRODUCTION

Vibrational spectroscopy is frequently used for the investigation of biological systems [1] because it provides detailed information about the local structure of the material. Most studies are carried out with infrared or Raman spectroscopies. With these techniques, the focus is on a few key bands, usually the amide I, (a coupled C=O stretch and N–H bend) the amide III, (a coupled C–N stretch and C–N–H in-plane bend) and, to a lesser extent, the X–H (X = C, N or O) stretch region. A complementary technique is inelastic neutron scattering (INS) spectroscopy [2]. The nature of INS spectra emphasises a 'whole spectrum' approach to vibrational analysis, the strengths and limitations of this method are discussed.

The major difference between vibrational neutron spectroscopy and infrared and Raman spectroscopies is that the neutron has mass, thus an inelastic scattering event results in a significant transfer of *both* energy (E, cm^{-1}) and momentum (Q, Å^{-1}). The energy transfer (E_T) is given by:

$$E_T = E_i - E_f \tag{1}$$

where the subscripts i and f refer to the incident and final values respectively. The momentum transfer is given by:

$$\underline{Q} = \underline{k}_i - \underline{k}_f \text{ where } k \equiv 2\pi / \lambda \tag{2}$$

k (Å^{-1}) is the neutron's wavevector; λ (Å) is the wavelength of the neutron.

Raman spectroscopy is also an inelastic scattering process, however in this case, the wavelength of both the incident and scattered radiation is several thousand Å, thus the wavevectors are very small so it follows that Q will also be very small. For infrared spectroscopy the incident wavevector is even smaller, since the wavelength is longer and for an absorption process, k_f is zero thus $Q = 0$. So both infrared and Raman spectroscopy are subject to the selection rule that only transitions at zero wavevector are observable. In contrast, INS spectroscopy is allowed for all wavevectors. Since both Q and E are experimentally and independently accessible, it follows that INS spectroscopy is intrinsically a two-dimensional form of spectroscopy.

INS spectroscopy can be either coherent, which can be thought of as inelastic diffraction, so gives information on collective motions of the system or incoherent which only involves the correlation between the position of the same nucleus, so there are no interference effects and the motions of a single particle are probed. Whether the scattering is primarily coherent or incoherent depends on the relative size of the coherent and incoherent scattering cross section's (σ) of the scattering nuclei. The cross section's are both element and isotope dependent, in particular the incoherent cross section of ^{1}H (hydrogen) is 80.3 barn (1 barn = 1×10^{-28} m^2), whereas that of most other elements is <5 barn. Furthermore, the incoherent cross section of ^{2}H (deuterium) is 5.6 barn so deuteration results in changes in both frequency and intensity.

For incoherent scattering the INS intensity, $S(Q, \omega_i)$, of the *i*th mode is proportional to [2]:

*Address correspondence to Stewart F. Parker:** ISIS Facility, STFC Rutherford Appleton Laboratory, Chilton, Didcot, Oxfordshire, OX11 0QX, United Kingdom; E-mail: stewart.parker@stfc.ac.uk

Salvatore Magazù and Federica Migliardo (Eds)

$$S(Q, \omega_i) \propto Q^2 U_i^2 \exp\left(-Q^2 U_{Total}^2\right)\sigma \qquad\qquad (3)$$

Where U_i is the amplitude of vibration of the atoms undergoing the particular mode, ω_i. Since the amplitude of vibration is greatest for ^1H and its incoherent cross section is at least a factor of ten times larger than that of most elements, for hydrogenous materials the scattering is dominated by hydrogen. The exponential term in equation (3) is the Debye-Waller factor, U_{Total} is the mean square displacement of the molecule and its magnitude is in part determined by the thermal motion of the molecule. This can be reduced by cooling the sample and so spectra are typically recorded below 30 K.

The key feature of Eq. (3) is that it is purely dynamic, the intensity does not depend on the electronic properties of the system. Thus any method that can provide the U_i values can be used to generate a calculated INS spectrum. Historically, the U_i values were obtained from 'ball and spring' models of molecules using the Wilson GF matrix method [3]. More recently, molecular mechanics has been used [4], molecular dynamics can also be used [5], however, for most systems *ab initio* methods [2,6-8] generally provide the best results.

Neutrons are highly penetrating, since the scattering event is between the neutron and the atomic nucleus, this means that sample presentation is straightforward, for air stable materials wrapping the sample in aluminium foil is often sufficient. Where samples are air sensitive or require controlled humidity, thin-walled sealed aluminium cans are used. A consequence of the high penetration is that scattering events are relatively rare, thus the sample size for INS spectroscopy is in the 0.5 – 5.0 g range and measurement times range from a few hours to 24 hours.

In the following sections, the advantages of INS spectroscopy are illustrated, complementarity with other spectroscopic techniques is demonstrated and the synergy with calculations is stressed.

COMPLEMENTARITY OF INS, INFRARED AND RAMAN SPECTROSCOPIES

Fig. **1** shows the infrared, Raman and INS spectra of L-cysteine [9]. As with all amino acids, this exists as a zwitterion in the solid state. Similarities and differences are apparent between all three types of spectra and emphasise the need to have all three types of spectra for a complete analysis. The similarities arise because the vibrational energy levels are an intrinsic property of the system, hence the energy differences between them (the vibrational transition energies) are independent of the method of observation. In contrast, the intensities depend on the technique used to observe the transition. The three forms of vibrational spectroscopy exploit different properties of the molecule. Infrared spectroscopy requires a change in dipole moment, thus is sensitive to the polar motions of the molecule, *e.g.* the asymmetric carboxylate stretch at 1580 cm^{-1}, while Raman spectroscopy requires a change in polarisability, thus is more sensitive to the non-polar motions of the molecule, *e.g.* the C–S stretch at 639 cm^{-1}.

In contrast, as a consequence of Eq. (3), the INS spectrum is dominated by the hydrogenic motions. The INS spectrum emphasises the modes that involve substantial hydrogen motion either directly (*e.g.* the CH$_2$ rock at 763 cm^{-1}) or where the hydrogen is "riding" on another atom's motion (*e.g.* the NH$_3$ torsion at 487 cm^{-1}).

The INS spectra of cysteine shown in Figs. **1c** and **1d** were recorded with two different INS instruments: MARI (1c) and TOSCA (1d) [2]. INS spectroscopy is a function of both energy transfer and momentum transfer. The design of TOSCA is such that there is a single momentum transfer associated with each energy transfer, *i.e.* TOSCA follows a fixed trajectory through (Q,ω) space. The consequences of this, are that the instrument is very simple to operate and that the resulting spectra are not dissimilar to infrared and Raman spectra, Fig. **1a,b**. MARI has a different operating principle and can measure both Q and ω independently. The price of this is increased complexity and that the incident energy, E_i, is fixed, thus it is only possible to measure transitions with an energy less than E_i. The Q variation is obtained by varying the scattering angle, so MARI has detectors at 3 - 130°. Fig. **2** shows a MARI plot and its relationship to the TOSCA plot: vertical lines of intensity correspond to peaks in the TOSCA plot. By examining the variation with Q at constant ω, it is possible to unambiguously determine whether a mode is a fundamental or an overtone [10].

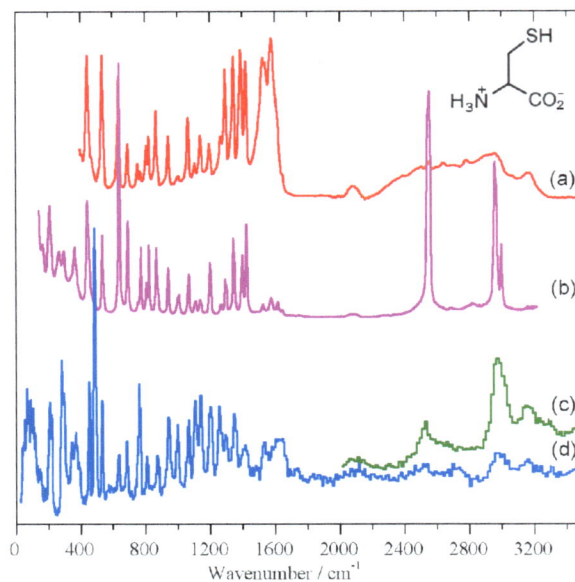

Figure 1: Vibrational spectra of L-cysteine. (a) infrared spectrum, (b) Raman spectrum, (c) INS spectrum recorded on MARI and (d) INS spectrum recorded on TOSCA. Reproduced from [9] with permission of IOS Press.

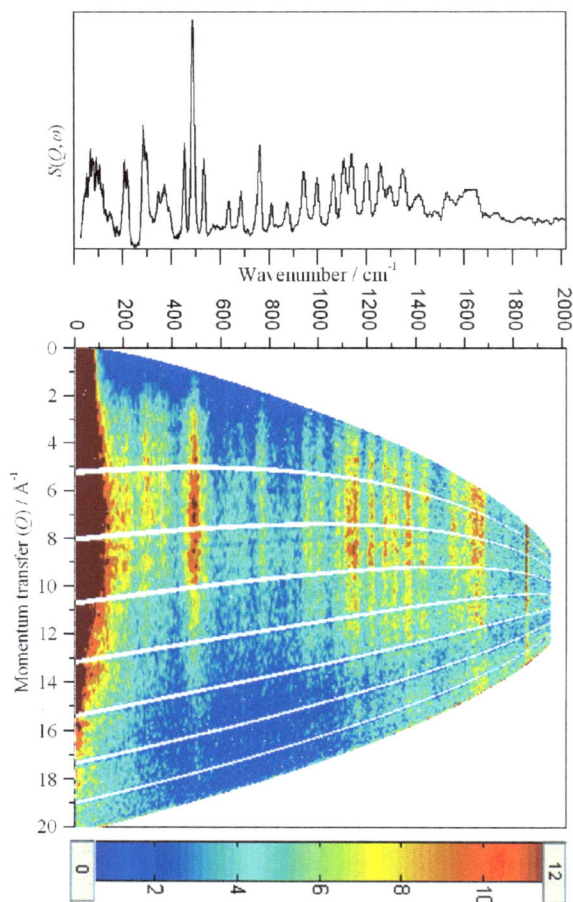

Figure 2: INS spectra of L-cysteine recorded with TOSCA (upper) and MARI (lower) and. Ridges of intensity in the MARI spectrum correspond to peaks in the TOSCA spectrum. Reproduced from [9] with permission of IOS Press.

A major difference between INS spectroscopy and infrared and Raman spectroscopies is that there are no selection rules and, in principle, all modes are observable. This arises because the interaction of the neutron is with the nucleus, not with the electrons as for the optical methods.

Combinations modes and overtones are allowed transitions in INS spectroscopy, whose intensity is proportional to Q^{2n}, where n = 1 for a fundamental, n = 2 for a binary combination or first overtone, n = 3 for a ternary combination or second overtone, *etc*… thus there are many more spectral features at higher energies. More importantly, however, spectral features become inherently broad at higher energy transfers due to the fact that significant momentum transfer takes place during the scattering process. Infrared and Raman spectroscopies, being essentially zero momentum transfer processes, do not suffer from this limitation and bands at high energy transfer are still sharp. It is worth noting that INS spectroscopy allows ready access to the low energy or terahertz region (< 400 cm^{-1}). While it is possible to obtain infrared and Raman spectra in this region, it becomes progressively more difficult to do so. Thus infrared spectra are routinely obtained down to 400 cm^{-1}, determined by the KBr cut-off. Operation below 400 cm^{-1} requires a change of beamsplitter and optics. Most Raman instruments employ a notch filter to eliminate the Rayleigh scatter, these typically cut-off in the 100 – 200 cm^{-1} range. To obtain spectra below this, requires at least a double monochromator with its attendant throughput penalty.

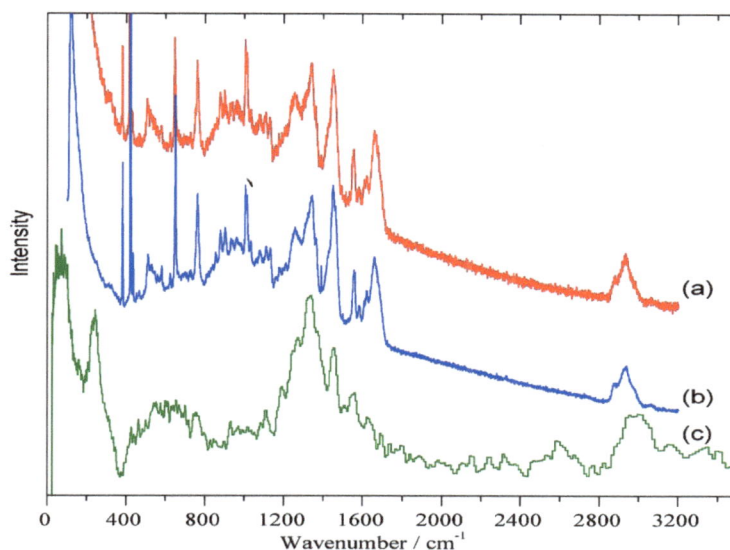

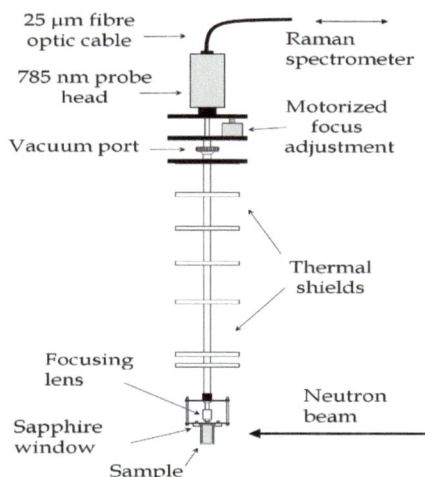

Figure 3: Raman spectra of lysozyme at (a) room temperature, (b) 20 K and (c) the INS spectrum at 20 K. Both the Raman and INS spectra were recorded simultaneously using the apparatus shown below. Reproduced from [11] with permission of the Society for Applied Spectroscopy.

The necessity to measure INS spectra at low temperature is often quoted as a reason why INS results are not relevant to biology. Proteins are known to cold denature and this is considered to invalidate the INS results. In an attempt to address this problem, the complementarity of INS and Raman spectroscopies has been exploited to enable simultaneous measurement of the spectra [11]. Fig. **3** shows the Raman spectrum of chicken egg white lysozyme at room temperature (3a), 20 K (3b) and the INS spectrum at 20 K (4c) recorded at the same time as the Raman spectrum. Clearly, Figs. **3a** and **3b** are very similar, apart from the band sharpening that typically occurs on cooling, indicating that the enzyme, as far as the spectroscopy is concerned, is unchanged on cooling, thus validating the conclusions from the INS spectra.

Comparison of Figs. **3b** and **3c** highlights the complementarity of the spectra, bands around 1400 cm^{-1} assigned to CH, CH$_2$ and CH$_3$ bending modes occur in both spectra, the carbonyl modes of the amide linkages are seen around 1700 cm^{1} in the Raman spectrum and the methyl torsions are seen as a broad band centred at 250 cm^{-1} in the INS spectrum. The width of this band (~50 cm^{-1}) is much greater than the resolution width at this energy (<5 cm^{-1}). This situation is very similar to that observed for the methyl groups in poly(methyl methacrylate) [12] where the spectra have been analyzed in terms of a model that considers a Gaussian distribution of potential barriers for methyl group rotation.

INS STUDIES OF PROTEINS

Proteins are molecular machines that are implicit in many important biological functions including enzymatic catalysis, molecular recognition, energy conversion and signal transduction. As proteins work largely in an aqueous environment at room temperature, it is natural to consider that dynamical or structural fluctuations are closely related to function. In order to understand the function, it is necessary to understand both the structure of the protein and the nature of the protein dynamics.

One of the great strengths of INS is that it readily provides a stringent test of theoretical models *via* Eq. (3). Fig. **4** shows a comparison [13] of the INS spectrum of Staphylococcal nuclease (SNase) with the labile protons replaced by deuterons and the spectrum derived from the molecular mechanics programme CHARMM. Most of the measured bands are present in the calculated spectrum. However, some clear differences between experiment and theory do exist. One is in the position of the methyl torsion at ~240 cm^{-1} which is found in the theoretical spectrum at ~276 cm^{-1}. Another difference is in the shape of the massifs at 1350 cm^{-1}. The peak at ~520 cm^{-1} in the experimental data is absent in the calculations. This is assigned to residual water in the lyophilized powder sample.

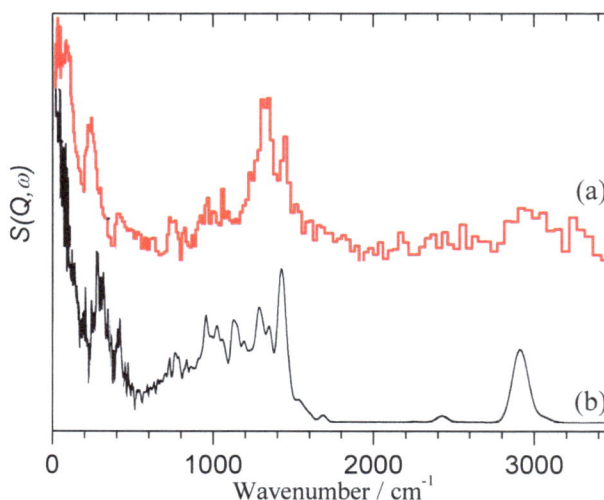

Figure 4: Observed (a) and calculated (b) INS spectrum of Staphylococcal nuclease. Reproduced from [13] with permission from the American Chemical Society.

Water molecules were not included in the normal mode analysis. The discrepancies between the spectra point to weaknesses in the model and highlight areas for improvement.

Examination of the displacement vectors of the modes giving rise to the calculated INS spectrum in Fig. **4** showed that most of the peaks below 1500 cm^{-1} do not arise from modes that can be simply described and contain contributions from many non-degenerate vibrations. However, it was possible to qualitatively identify some common features in the displacements contributing to many of the theoretical peaks. The assignments generally follow those found in organic molecules: 350 – 500 cm^{-1} skeletal bending modes and torsions, 700 – 1000 cm^{-1} C–C stretch, CH$_2$ and CH$_3$ rock, 1000 – 1500 cm^{-1} CH$_2$ rock, wag, twist, scissors and CH$_3$ symmetric and asymmetric bend.

CHARMM has also been used to analyse the low energy modes of DNA [14] observed by coherent INS and by inelastic X-ray scattering. Base-pair opening plays a key role in replication, transcription and denaturation. These processes all involve the splitting of the double helix into single strands and while they involve proteins interacting with DNA, they are thought to be driven by the dynamics of DNA itself. In particular, a strongly dispersive mode with a maximum energy of ~100 cm^{-1} has been extensively studied. The upper part of Fig. **5** shows the dispersion relations for the first 20 modes in the direction of the helix axis. The acoustic phonons have a maximum energy of 10 cm^{-1} at the Brillouin zone boundary. Above this the optic modes have limited dispersion. At the gamma point of the Brillouin zone, the optic modes start at 13.7 cm^{-1} and there are ~1000 modes in the first 100 cm^{-1}. At 100 cm^{-1} the vibrational modes are already significantly localized, which means there is no coherent base pair opening over a significant length of the helix.

Calculating the spectral intensities for all modes for coherent inelastic X-ray scattering and INS for wave vectors up to 3 Å^{-1} gives total spectral profiles. These are well-fitted with a simple Gaussian, which enables the characteristic transition energy and width of the spectra to be determined for each wave vector.

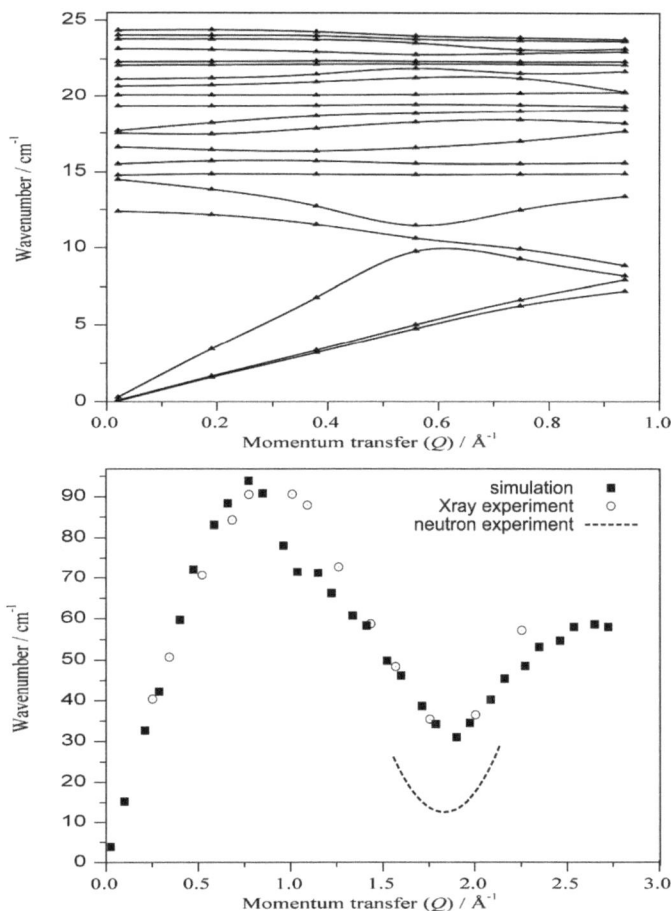

Figure 5: (Upper) Calculated dispersion curves of the 20 lowest modes of DNA and (lower) the apparent dispersion obtained by fitting the spectral intensity. Reproduced from [14] with permission from the American Physical Society.

The dispersion curve obtained in this way, lower part of Fig. **5**, agrees well with the experimental data in terms of transition energy and width, the width being comparable to the transition energy. The experimental signal is not a direct measurement of the acoustic phonon but it is the projection of spectral intensity over a large number of mainly optic modes [14].

Protein dynamics are closely related to the function and the unique tertiary structure is required to express function. The questions arises as to whether there are specific spectral features characteristic of the folded native protein and whether these features property vary from protein to protein. SNase is very useful in this respect because there is a mutant form that lacks 13 residues from its C-terminus which exists in a partly unfolded form. Fig. **6** compares the INS spectra of the folded and partly unfolded forms [15]. Careful inspection reveals some differences: the peak at ~150 cm^{-1} is only present in the folded state, while peak at about 770 cm^{-1} is only present in the unfolded state. The differences probably reflect different secondary structures. Differences in tertiary structure should be more apparent in the region below 100 cm^{-1}. Comparison of the INS spectra [16] of SNase, the SNase mutant and myoglobin showed no significant differences between the three spectra, despite the significant differences in the tertiary structures. (Myoglobin is a typical α-protein, while SNase contains three α-helices and a core composed by five β-sheets. The SNase fragment is in a compact denatured state, where the α-helical components are almost lost). A peak at ~30 cm^{-1} was shown to shift to lower energy with increasing molecular weight of the protein. This behaviour is similar to that found for the Boson peak in polymers and suggests that the low frequency modes are extended over the whole molecule, *i.e.* in terms of the low energy dynamics, the protein can be regarded as a continuous elastic body.

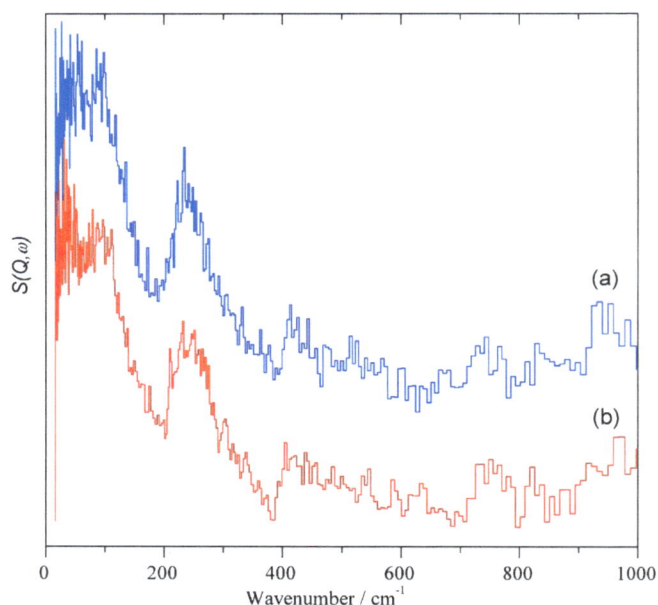

Figure 6: Comparison of the INS spectra between (a) wild type SNase (folded form) and (b) the truncated mutant (unfolded form). Reproduced from [16] with permission of IOS Press.

Collagen is a fibrous protein and is the most abundant protein in mammals. It is tough and inextensible, with great tensile strength and is the main component of cartilage, ligaments and tendons, bone and teeth. Type I collagen is the most abundant collagen in the human body and is constituted almost entirely of three chains of amino acids, each chain having the repeat glycine–proline–hydroxyproline (Gly–Pro–Hyp). These three chains are wound together in a tight triple helix through hydrogen-bonds perpendicular to the helix axis. The biomechanical properties of collagen depend on its triple helix structure and the corresponding low frequency vibrations.

The polypeptides poly-glycine-II (PG-II), poly-proline-II (PP-II) and the synthetic polypeptide (Pro–Pro–Gly)$_{10}$ (PPG)$_{10}$ have been studied by INS as model compounds of type I collagen [17,18]. Fig. **7** shows the experimental INS spectra and those calculated [18] by periodic-DFT for PG-II, PP-II and (PPG)$_{10}$. For PG-II and PP-II the agreement is generally good. However, for $(PPG)_{10}$, multiple imaginary modes are found below 100 cm^{-1}. The flexibility of the helical structure, the lack of symmetry, the large number of degrees of freedom and the partially amorphous nature of $(PPG)_{10}$, mean that there are many minima of similar energy which renders finding the ground state extremely difficult. This is likely to be a significant problem, even when in the near future computational methods allow thousands of atoms to be treated, meaning that small proteins become feasible. The parameterised force field methods do not suffer from this problem and were successfully used to model the low energy region of $(PPG)_{10}$ [18].

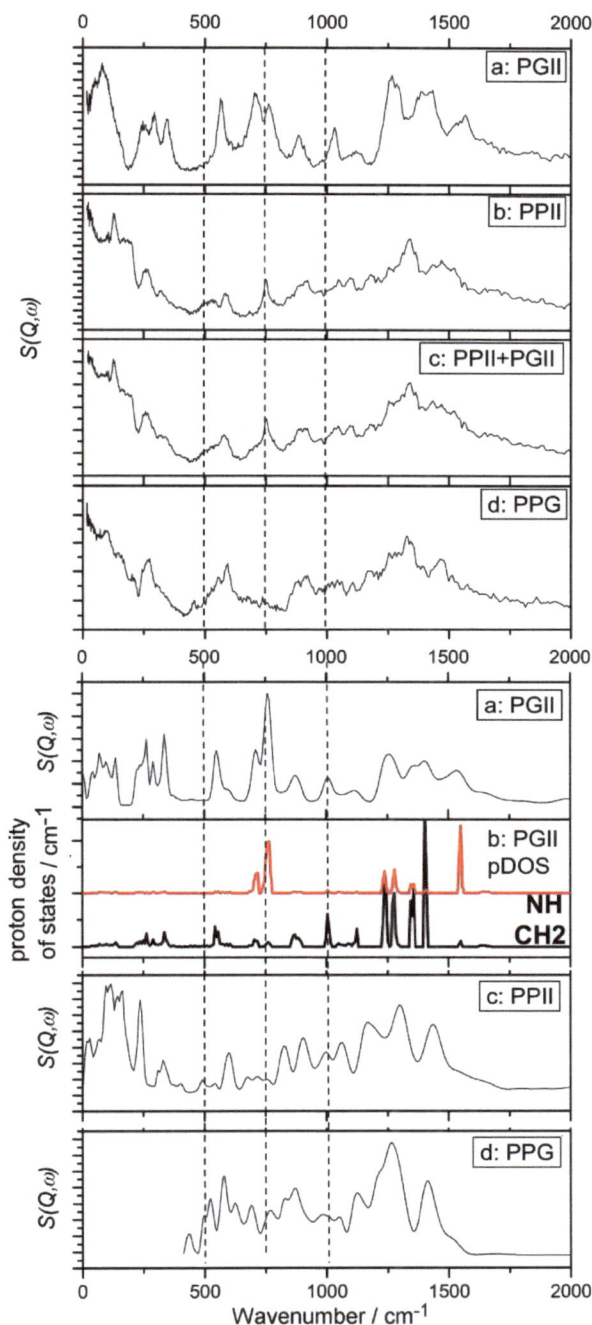

Figure 7: Upper: Experimental INS spectra of $(PPG)_{10}$ and the component polypeptides. Panel (c) is the proton and residue weighted sum of the PG-II and PP-II spectra. Lower: Calculated INS spectra of $(PPG)_{10}$ and component polypeptides. Panel (b) shows the partial densities of states for the different protons in PG-II. Reproduced from [18] with permission from Elsevier.

The acoustic phonons show dispersion along the helix axis and reach a maximum value of 15 cm^{-1} at the Brillouin zone boundary of the helix. The torsional character of modes is pronounced between 0 and 60 cm^{-1} and at some higher frequencies, such as 140 cm^{-1}. There is only one mode with strong, pure rotation character at 20 cm^{-1}. Modes with breathing character (out-of-phase radial motion of the three amino acids which belong to the same plane perpendicular to the triple helices axis) are most evident between 60 and 120 cm^{-1}, with a maximum at ~100 cm^{-1}. The higher energy modes, which are more localized are less well described and this clearly highlights the utility of INS spectra as a test bed for the refinement of such methods.

WATER

Water may seem an odd topic for a review of biomacromolecular systems, however, in biology, hydration is a very important factor for stability and function. A number of studies strongly suggest that the dynamical transition of proteins, which is the transition from a low temperature harmonic regime to an anharmonic one at higher temperature, is intimately linked to the solvent dynamics. This transition is a direct consequence of the large amplitude, anharmonic, atomic motions of the protein and these motions, necessary for biological activity, are highly dependent on the degree of plasticizing, which is determined by the level of hydration.

Water, as ice, has been extensively studied by INS spectroscopy [19,20]. Even nominally dry proteins or model compounds contain substantial amounts of water. Thus 'dry' (PPG)$_{10}$ retains one water molecule per PPG unit and has a water content of 7 (g H$_2$O)(100 g anhydrous (PPG)$_{10}$)$^{-1}$, comparable to that found in 'dry collagen of 6 (g H$_2$O)(100 g anhydrous collagen)$^{-1}$. This water is easily detectable either by its ready exchange with D$_2$O as shown in Fig. **8** (upper part) or by comparison of material at different levels of hydration, Fig. **8** (lower part) [17].

The difference spectrum [(25% hydrated collagen) – (dry collagen)], Fig. **8** (lower part) reveals additional intensity, in the hydrated sample, in the acoustic phonon region around 50 cm^{-1}. This corresponds to an acoustic band observed in pure H$_2$O ice [19,20]. At 25% hydration all the water in the collagen sample is closely associated with protein and there is no pure ice. The collagen helices pack in a quasihexagonal arrangement and water molecules form extended, hydrogen-bonded chains and clusters. Such extended networks are capable of sustaining collective phonon excitations and the 50 cm^{-1} peak in the difference spectrum is assigned to phonon excitations propagating through this interhelical water fraction. This assignment is supported by the absence of a similar 50 cm^{-1} peak in the difference spectrum between hydrogenous and deuterated (PPG)$_{10}$, Fig. **8** (upper part). The 7 wt% water present in the (PPG)$_{10}$ samples is in tightly bound sites on the peptide backbone. At this hydration level, water does not form an extended hydrogen-bonded network and so cannot sustain collective ice-like phonon excitations. In contrast, D$_2$O exchange of the tightly bound water fraction in (PPG)$_{10}$ produces a marked loss of intensity in a broad band between 450 and 800 cm^{-1}, as revealed in the difference spectrum. This corresponds to a similar loss of intensity between 450 and 800 cm^{-1} on drying of 25% H$_2$O hydrated collagen. The detailed features of the two difference spectra are not identical. The collagen difference spectrum shows intensity extending to 1100 cm^{-1}, with further broad differences above 1500 cm^{-1}, suggesting that the water interacts with many parts of the collagen backbone, rather than the specific, localized interaction found in (PPG)$_{10}$.

There is a growing body of evidence that the structure and behaviour of water in the vicinity of biological macromolecules may be different from that of pure water. High resolution crystal structures of pure protein or DNA that have been obtained at cryogenic temperatures have ordered interfacial water molecules in the electron density maps, and in a few cases these represent a large fraction of the total solvent content in the crystals. However, these ordered water molecules do not form the typical ice I_h structure, but rather are involved in many different forms of hydrogen bonding networks with the macromolecule and with each other. If such an altered water structure exists in the vicinity of macromolecules in cells, then it has potential significance for a range of fundamental functions such as protein folding and stability, DNA packaging, and molecular recognition. Moreover, an understanding of the interaction of water and macromolecules at cryogenic temperatures is, in its own right, of considerable importance, especially for the freezing of cells and tissues without the irretrievable loss of function. As was done in the case of collagen [17], differences in the INS spectra as a function of hydration can be used to investigate how the water is modified by its interaction with the protein.

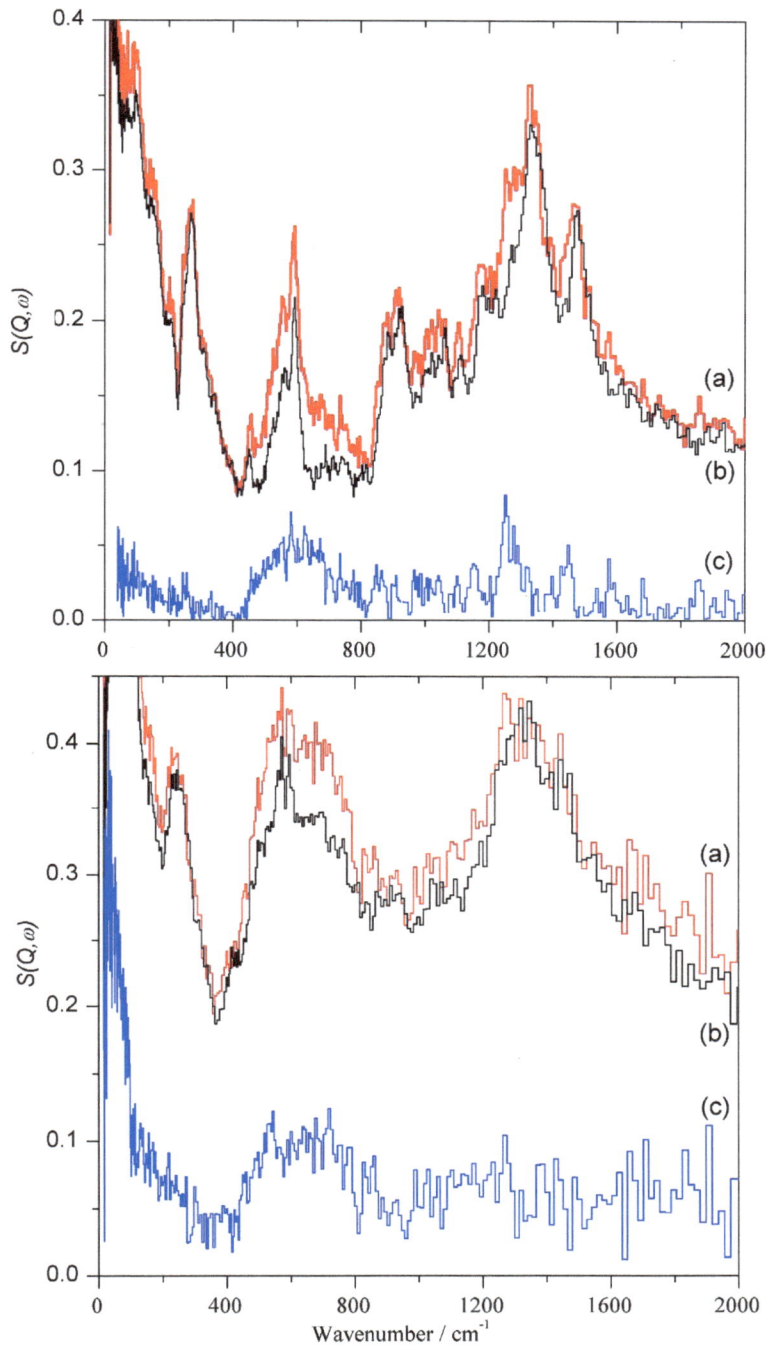

Figure 8: Upper: INS spectra, $(PPG)_{10}$-H (a, red line), and $(PPG)_{10}$-D (b, black line) and the difference (H − D) (c, blue line). Lower: INS spectra of collagen, 25% hydrated (a, red line) and 'dry' (b, 6% water) (black line) and the difference (25% − 6%) (c, blue line). Reproduced from [17] with permission from Elsevier.

This has been done for a variety of materials including grana membranes isolated from spinach [21], DNA [21] and intact cells from bovine heart muscle, beetroot, yeast, spinach and green algae (*Chlamydomonas reinhardtii*) [22]. Fig. **9** shows a series of spectra of DNA at different hydrations. Bulk water (*i.e.* ice I_h) is readily identified by the vertical 'cliff edge' at 545 cm^{-1} and this signal is not detected below a certain water concentration, in this case ~50 g H_2O (100 g dry DNA)$^{-1}$. At concentrations of water above this point, bulk water can be detected, although the spectra indicate that interfacial water continues to accumulate toward a saturation level, as shown in Fig. **10** (upper part).

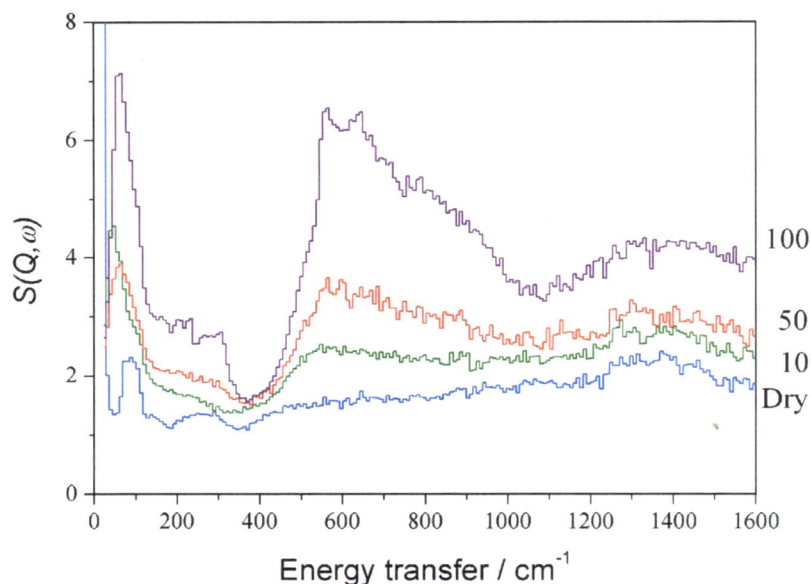

Figure 9: INS spectra for DNA at increasing levels of hydration corresponding to 0 (dry), 10, 50, and 100 g of water per 100 g of dry DNA. Reproduced from [21] with permission from the American Chemical Society.

At lower concentrations, the vibrational spectra indicate that all of the water molecules are perturbed, see Fig. **10** (lower part).

By comparison to ice I_h, it can be seen that the librational edge is shifted down in energy and is much less distinct. The sharp features due to optic translational modes in ice I_h at 227 and 302 cm^{-1} are absent. Together, the data imply that the interfacial water is strongly disordered with less hydrogen-bonding than ice I_h.

CONCLUSIONS

INS spectroscopy of biomacromolecules is difficult. The usual advantages of INS: absence of selection rules, sensitivity to hydrogen motion, not restricted to gamma point modes and high penetration, mean that biopolymers with hundreds, if not thousands, of hydrogen atoms yield dense, highly congested spectra. However, the ability to compare calculated and experimental spectra is a major strength of INS spectroscopy since it provides a very stringent test of the calculation, in a way that comparison of structure alone cannot. For small (up to a few hundred atoms) periodic-DFT provides a uniquely powerful method to understand the systems. For large systems with less well-defined ground states, it is clear that force field methods still have a significant role to play. By providing a test bed for the empirical potentials and force constants employed, INS spectra can continue to make a useful contribution to this field.

At present, since relatively large quantities of sample are needed ~1 g, studies are generally restricted to commonly available materials such as DNA, lysozyme and collagen. With the advent of new high power neutron sources (TS-II at ISIS, UK [23], SNS at Oak Ridge, USA [24] and J-PARC at Tokai, Japan [25]) with improved instrumentation becoming available in 2010-2015, the sample size required will certainly decrease to ~100 mg and perhaps 10 mg or so. These sample sizes are more compatible with standard preparative routes in biology so will expand the utility of INS for studies of biomacromolecules.

ACKNOWLEDGEMENTS

The STFC Rutherford Appleton Laboratory is thanked for access to neutron beam facilities. Dr Mark R. Johnson (Institut Laue-Langevin) is thanked for providing Figs. **5** and **7**.

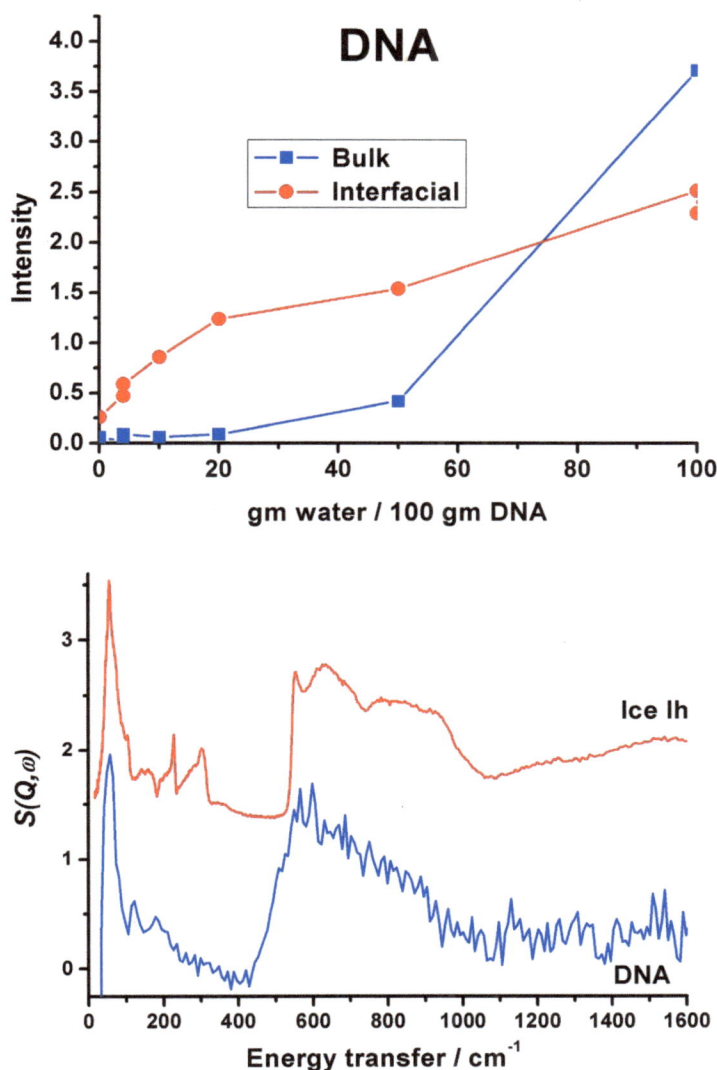

Figrue 10: Upper: Bulk water and interfacial water signal magnitudes as a function of hydration of DNA. Bulk water (boxes) is measured by the sharp rise at 532-556 cm^{-1}, while interfacial water (circles) is given by the magnitude of the broad increase from 400 to 589 cm^{-1} (from which any bulk signal is subtracted). Lower: Comparison of the INS spectra of ice I_h (upper) and an estimation of the INS spectrum of interfacial water in hydrated DNA (50 g of water/100 g of DNA) obtained by subtraction of the dry DNA component. Reproduced from [21] with permission from the American Chemical Society.

REFERENCES

[1]   Schweitzer-Stenner R. Advances in vibrational spectroscopy as a sensitive probe of peptide and protein structure: A critical review. Vib Spec 2006; 42: 98-117.

[2]   Mitchell PCH, Parker SF, Ramirez-Cuesta AJ, Tomkinson J. Vibrational Spectroscopy with Neutrons, with Applications in Chemistry, Biology, Materials Science and Catalysis. Singapore, World Scientific, 2005.

[3]   Wilson EB Jr, Decius JC, Cross PC. Molecular Vibrations. Dover, New York 1955.

[4]   Goupil-Lamy AV, Smith JC, Yunoki J, Tokunaga F, Parker SF, Kataoka M. High-resolution vibrational inelastic neutron scattering: A new spectroscopic tool for globular proteins. J Amer Chem Soc 1997; 119: 9268-73.

[5]   Ramirez-Cuesta AJ, Mitchell PCH, Parker SF, Wilkinson AP, Rodger PM. Molecular dynamics simulation of inelastic neutron scattering spectra of librational modes of water molecules in a layered aluminophosphate. In: Johnson MR, Kearley GJ, Büttner HG. Eds. Neutrons and Numerical Methods-N(2)M. New York, AIP Conf. Proc. 1999, Vol.479,

Ch.37, pp.195-200.

[6]   Johnson MR, Parlinski K, Natkaniec I, Hudson BS. Ab initio calculations and INS measurements of phonons and molecular vibrations in a model peptide compound – urea. Chem Phys 2003; 291: 53–60.

[7]   Parker SF, Jeans R, Devonshire R. Inelastic neutron scattering, Raman spectroscopy and periodic DFT study of purine. Vib. Spec. 2004; 35: 173-7.

[8]   Hudson BS. Vibrational spectroscopy *via* inelastic neutron scattering. In Laane J. Ed. Frontiers of Molecular Spectroscopy. Elsevier, 2009.

[9]   Parker SF, Haris P. Inelastic neutron scattering spectroscopy of amino acids. Spectroscopy Int J 2008; 22: 297-307.

[10]   Parker SF, Bennington SM, Ramirez-Cuesta AJ, Auffermann G, Bronger W, Herman H, Williams KPJ, Smith T. Inelastic neutron scattering, Raman spectroscopy and periodic-DFT studies of Rb_2PtH_6 and Rb_2PtD_6. J Amer Chem Soc 2003; 125: 11656-61.

[11]   Adams MA, Parker SF, Fernandez-Alonso F, Cutler DJ, Hodges C. King A. Simultaneous neutron scattering and Raman scattering. Appl Spec 2009; 63: 727-32.

[12]   Moreno AJ, Alegrý A, Colmenero J, Frick B. Methyl group dynamics in poly(methyl methacrylate): from quantum tunnelling to classical hopping. Macromolecules 2001; 34: 4886-96.

[13]   Goupil-Lamy AV, Smith JC, Yunoki J, Parker SF, Kataoka M. High-resolution vibrational inelastic neutron scattering: A new spectroscopic tool for globular proteins, J Amer Chem Soc 1997; 119: 9268–73.

[14]   Merzel F, Fontaine-Vive F, Johnson MR, Kearley GJ. Atomistic model of DNA: Phonons and base-pair opening. Phys Rev E 2007; 76: 031917-22.

[15]   Kataoka M, Kamikubo H, Nakagawa H, Parker SF, Smith JC. Neutron inelastic scattering as a high-resolution vibrational spectroscopy: New tool for the study of protein dynamics. Spectroscopy Int J 2003; 17: 529-35.

[16]   Kataoka M, Kamikubo H, Yunoki J, Tokunaga F, Kanaya T, Izumi Y, Shibata K. Low energy dynamics of globular proteins studied by inelastic neutron scattering. J Phys Chem Solid 1999; 60: 1285–9.

[17]   Middendorf H.D, Hayward RL, Parker SF, Bradshaw J, Miller A. Vibrational neutron spectroscopy of collagen and model polypeptides. Biophysical Journal 1995; 69: 660-73.

[18]   Fontaine-Vive F, Merzel F, Johnson MR, Kearley GJ. Collagen and component polypeptides: low frequency and amide vibrations. Chem Phys 2009; 355: 141–8.

[19]   Li J-C, Londono JD, Ross DK, Finney JL, Sherman WF, Tomkinson J. An inelastic incoherent neutron scattering study of ice II, IX, V and VI - in the range from 2 to 140 meV. J Chem Phys 1991; 94: 6770-5.

[20]   Li J-C. Inelastic neutron scattering studies of hydrogen bonding in ices. J Chem Phys 1996; 105: 6733-55.

[21]   Ruffle SV, Michalarias I, Li J-C, Ford RC. Inelastic incoherent neutron scattering studies of water interacting with biological macromolecules. J Amer Chem Soc 2002; 124: 565-9.

[22]   Ford RC, Ruffle SV, Ramirez-Cuesta AJ, Michalarias I, Beta I, Miller A, Li J-C. Inelastic incoherent neutron scattering measurements of intact cells and tissues and detection of interfacial water. J Amer Chem Soc 2004; 126: 4682-8.

[23]   http://www.isis.stfc.ac.uk/

[24]   http://www.sns.gov/

[25]   http://j-parc.jp/MatLife/en/index.html

CHAPTER 8

Integration of Neutron Scattering with Computer Simulation to Study the Structure and Dynamics of Biological Systems

Jeremy C. Smith*, Marimuthu Krishnan, Loukas Petridis and Nikolai Smolin

UT/ORNL Center for Molecular Biophysics, Oak Ridge National Laboratory, Oak Ridge, TN 37831, USA

Abstract: A synergistic approach is described combining computer simulation with experiment in the interpretation of small angle neutron scattering (SANS) and inelastic scattering experiments on the structure and dynamics of proteins and other biopolymers. Simulation models can be tested by calculating neutron scattering structure factors and comparing the results directly with experimental scattering profiles. If the scattering profiles agree the simulations can be used to provide a detailed decomposition and interpretation of the experiments, and if not, the models can be rationally adjusted. Comparison with neutron experiment can be made at the level of the scattering functions or, less directly, of structural and dynamical quantities derived from them. This methodology is discussed in the context of the protein glass transition, protein-solvent dynamical coupling, the density of the hydration shell of proteins and the structural analysis of lignocellulosic biomass.

INTRODUCTION

The characterization of the structure and internal dynamics of biomolecules such as proteins is essential to understanding the mechanisms of their biological functions. In this regard the combination of molecular dynamics simulation and neutron scattering techniques has emerged as a highly synergistic approach [1]. Neutron scattering can be used to test molecular simulation models by direct comparison in two ways. The first involves directly comparing experimental and calculated scattering intensities. Secondly, one can indirectly compare experiment and simulation by examining "derived" quantities, such as, for example, the radius of gyration and the fractal dimension of a biomolecules. However, obtaining these derived quantities from both experiment and simulation requires approximations and model-dependent data interpretation.

As well as comparison with experiment, computer simulation can be used in the theoretical interpretation of experimental data. For example, in small-angle neutron scattering (SANS) heterogeneous, multi-component biological systems produce complex scattering patterns that can be difficult to interpret analytically, especially when the scattering length densities of the different components are similar. This problem is circumvented in part experimentally by the use of contrast variation techniques that make it possible to separate scattering that results from the different components through the controlled replacement of hydrogen with deuterium. However, the complex task of identifying scattering contributions from the various components is further simplified with the use of computer simulation, and especially molecular dynamics (MD). Once MD simulation has been performed, it is straightforward to compute scattering intensities of the whole system or of the individual components. In other words, MD simulation can be considered as a "*in silico* contrast variation" technique.

This chapter provides an overview of combined applications of neutron scattering experiments and computer simulations to understand molecular motions of proteins and biopolymers. Specifically, the examples gives overview of developments in the applications of dynamic neutron scattering to understand the protein glass transition and role of solvent interactions in controlling protein dynamics, and SANS to develop atomic resolution models of large biomolecular complexes.

NEUTRON SCATTERING FUNCTIONS AND ATOMIC FLUCTUATIONS

This chapter begins with a brief introduction to some basic equations relating neutron scattering to structure and dynamics that are relevant to computer simulation.

In biological neutron scattering experiments, neutron beams of suitable wavelengths are scattered by atomic nuclei of the sample, resulting in changes in the energy and momentum of the incident neutrons. By measuring these changes, information about the structure and internal dynamics of atoms/molecules that make up the sample can be inferred [2].

*Address correspondence to Jeremy C. Smith:** UT/ORNL Center for Molecular Biophysics, Oak Ridge National Laboratory, Oak Ridge, TN 37831, USA; E-mail: smithjc@ornl.gov

Salvatore Magazù and Federica Migliardo (Eds)

The measured quantity in dynamic scattering is the number of neutrons scattered within a solid angle between Ω and $\Omega+d\Omega$ with a change in energy $\hbar\omega$ and momentum $\hbar Q$. This number is proportional to the double-differential cross-section $\delta^2\sigma/\delta\Omega\delta\omega$, which in turn is proportional to the dynamic structure factor, $S(Q,\omega)$, [2]:

$$\frac{\partial^2 \sigma}{\partial\Omega\partial\omega} \; \alpha \;\; S(Q,\omega),$$

(1)

The dynamic structure factor can be written in terms of the van Hove function, $G(r,t)$, which characterizes the space-time correlation of individual atoms as well as between pairs of atoms, as shown in Fig. **1**. It is evident from Fig. **1** that Fourier transformation of physical quantities determining atomic dynamics in the "Molecular Dynamics Space" leads to information determined in "Neutron Scattering Space" and vice versa.

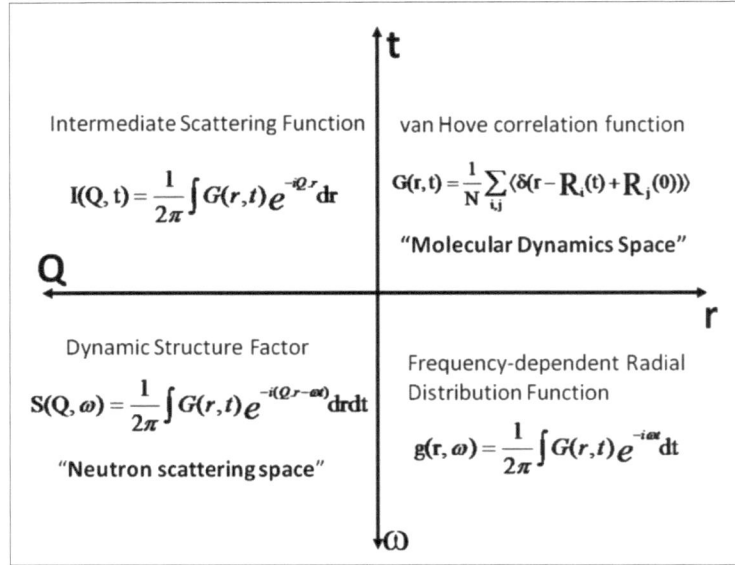

Figure 1: Schematic representation of the relationship between neutron scattering functions (determined from experiments) and the space-time correlation function (calculated using MD simulations) is shown here. Here, $R_i(t)$ denotes the atomic coordinates of ith atom at time t.

SMALL ANGLE NEUTRON SCATTERING

In Small Angle Neutron Scattering (SANS) experiments one measures the flux of neutrons of wavelength λ scattered into an angle θ. The scattering intensity $I(Q)$ is a function of the scattering vector

$$I(Q) \propto P(Q) \cdot S(Q), \;\; Q = \frac{2\pi}{\lambda}\sin\left(\frac{\theta}{2}\right),$$

(2)

where $P(Q)$ is the particle form factor and $S(Q)$ the particle structure factor. The proportionality sign in Equation 2 accounts for instrument-specific factors. In this work we will focus on the coherent scattering structure factor $S(Q)$ that is given by:

$$S(Q) = 4\pi\rho\int r^2\left[g(r)-1\right]\frac{\sin(Qr)}{Qr}dr,$$

(3)

where $g(r)$ is the pair-correlation function and ρ is the density of the scattering particles. The structure factor describes how $I(Q)$ is modulated by interference effects between radiation scattered by different scattering bodies and can be used to gain information about the relative positions of these scattering bodies.

APPLICATIONS OF NEUTRON SCATTERING AND MD SIMULATIONS TO BIOMOLECULAR DYNAMICS

Protein Glass Transition

Neutron scattering has been a major technique in protein glass transition research. Following the first neutron scattering experiment reporting glass transition behavior on hydrated myoglobin powders [3,4], it was demonstrated that the same transition is present in protein MD simulation [5]. Subsequently, a large number of neutron scattering and other experimental and computer simulation studies on various biological systems have revealed that many proteins exhibit this temperature-dependent dynamical transition around 180-220K [3,4,6-22]. Below this transition, the dynamics is similar to that of a glassy material, while at temperatures above ~220K, protein atoms exhibit liquid-like dynamics [23].

The evidence so far points to the glass transition being a general phenomenon among proteins. Furthermore, in some proteins correlations have been observed between the onset of protein function, such as ligand binding or proton pumping and the onset of the transition *i.e.*, it has been suggested that these proteins function only when the temperature is above the dynamical transition, although enzyme function below the transition has been demonstrated [14]. The 180-220K transition is sensitive to changes in solvent conditions. For example, there is evidence that proteins immersed in viscous solvents, such as trehalose, exhibit no transition [24].

The above observations led to the following questions concerning the microscopic dynamical details of the protein glass transition: (a) Is the dynamical transition in a solvated protein controlled by the solvent or does the intrinsic anharmonicity of protein dynamics also play a role? (b) Do proteins exhibit intrinsic anharmonic dynamics below the glass transition temperature?

The change in gradient of mean-square displacement (MSD) versus temperature is consistent with the dynamics of a protein changing from harmonic to anharmonic dynamics across the transition. However, neutron scattering experiments and molecular dynamics simulations have shown signatures of anharmonic dynamics well below the ~220K dynamical transition [15,20-22]. The dynamic processes associated with this low-temperature anharmonicity, and how these motions may be related to global dynamical changes at the dynamical transition, have yet to be fully understood.

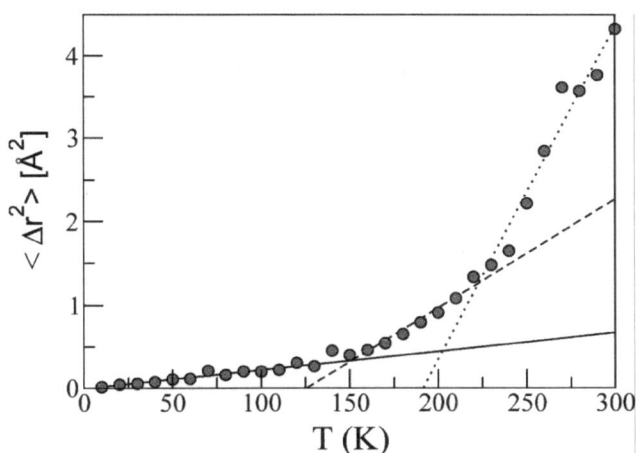

Figure 2: Time-averaged mean square displacement of myoglobin as a function of temperature from MD. The straight lines are fits to the data for different temperature ranges (solid line, 0 to 150 K; dashed line, 150 to 220 K; dotted line, above 220 K) and are shown as a guide to the eye. (Reproduced with permission from [26]. Copyright 2008 Am. Chem. Soc.)

A recent neutron scattering study on hen egg-white lysozyme showed the existence of a low-temperature onset of anharmonicity at around 100K, the origin of which was suggested to be methyl group rotation [21,22]. Indeed, in neutron scattering experiments, the main contribution to the scattered protein intensity arises from the nonexchangeable hydrogen atoms, and significant fraction of nonexchangeable hydrogens in proteins resides on CH_3 groups: 26% in lysozyme, for example. Thus, the CH_3 groups contribute significantly to the scattered intensity. Also, it has been suggested that a dominant contribution of the relaxation observed in dry myoglobin neutron scattering is due to methyl dynamics [10]. ^{1}H NMR relaxation studies have also investigated the reorientational dynamics of C-H bond vectors of methyl groups, and ^{1}H NMR experiments on dry lysozyme have shown that 70% of the total proton relaxation is due to methyl dynamics [25].

The average MSD of atoms of hydrated myoglobin as a function of temperature calculated from recent MD simulations is shown in Fig. **2** [26].

The MSD increases linearly at low temperatures then exhibits two slope changes: one at ~150 K and the other at ~220 K. The change at ~220 K is the solvent-driven dynamical transition as observed in many biological systems. At 150K, rotational excitations of methyl groups were observed in the MD and these jump-like motions of methyl protons will lead to quasi-elastic neutron scattering of the type that has been observed experimentally at ~150K in proteins [21,22, 26]. In lysozyme, the low-temperature anharmonicity was observed at 100 K and was attributed to the onset of methyl dynamics. It was also demonstrated that the anharmonic dynamics observed at ~100 K is independent of hydration level, while the dynamical transition at ~200-220K is observed only at hydration levels greater than 0.2 g water/g protein [27]. Thus, recent neutron scattering experiments and MD simulations have demonstrated the non-negligible role of intrinsic anharmonicity of protein dynamics in the protein glass transition.

Recently, the relationship between enzyme dynamics and activity at low hydration was examined [15,28,29]. It was found that significant intraprotein quasielastic scattering exists even below the dynamical transition. Furthermore, measurements have demonstrated enzyme activity at hydrations as low as 3% [29].

Finally, we draw attention to work designed to derive simplified analytical models for diffusive protein dynamics at 300 K. Molecular dynamics simulation of oligopeptide chains reveals configurational subdiffusion at equilibrium extending from 10^{-12} to 10^{-8} s. Trap models, involving a random walk with a distribution of waiting times, cannot account for the subdiffusion, which has found rather to arise from the fractal-like structure of the accessible configuration space [30,31]. These conceptual approaches will hopefully be of use in analysing quasielastic scattering.

Protein-Solvent Dynamical Coupling

Water plays a crucial role in determining the structures, dynamics and function of biomolecules. Water molecules in the hydration layer of biomolecules (biological water) are important not only for the thermodynamic stability of proteins, but also play a central role in several biomolecular functionalities, such as interaction, catalysis, recognition, etc. A number of experiments and simulations have indicated that the ~ 200 K dynamical transition is strongly coupled to the solvent [11,24,32-36].

To determine the driving force behind the protein glass transition, a set of molecular dynamics simulations of myoglobin surrounded by a shell of water was performed using a dual heat bath method, in which the protein and solvent are held at different temperatures [18]. The results show that the protein transition is driven by a dynamical transition in the hydration water that induces increased fluctuations primarily in sidechains in the external regions of the protein. The water transition involves activation of translational diffusion and occurs even in simulations where the protein atoms are held fixed [18,37].

Fig. **3a** presents the protein fluctuations calculated from a control set of simulations (in which in each simulation the protein and solvent are at the same temperature), together with those obtained by fixing the temperature of one component at a temperature below the dynamical transition while varying the temperature of the other. In the control set, the experimentally-known dynamical transition is reproduced, with nonlinearity starting at 220 K. Fixing the solvent temperature at 80 K or 180 K suppresses the dynamical transition, the protein MSD increasing linearly with temperature up to 300 K. Therefore, low temperature solvent cages the protein dynamics.

Fig. **3a** also shows that holding the protein temperature constant at 80 K or 180 K and varying the solvent temperature also abolishes the dynamical transition behavior in the protein. In summary, then, Fig. **3a** demonstrates that holding either component at a low temperature suppresses the protein dynamical transition.

Fig. **3,b** and **c**, shows the effect of holding one component above the transition temperature while varying the temperature of the other. Holding the solvent temperature at 300 K (Fig. **3b**) leads to increased protein fluctuations at most temperatures relative to the other simulation sets.

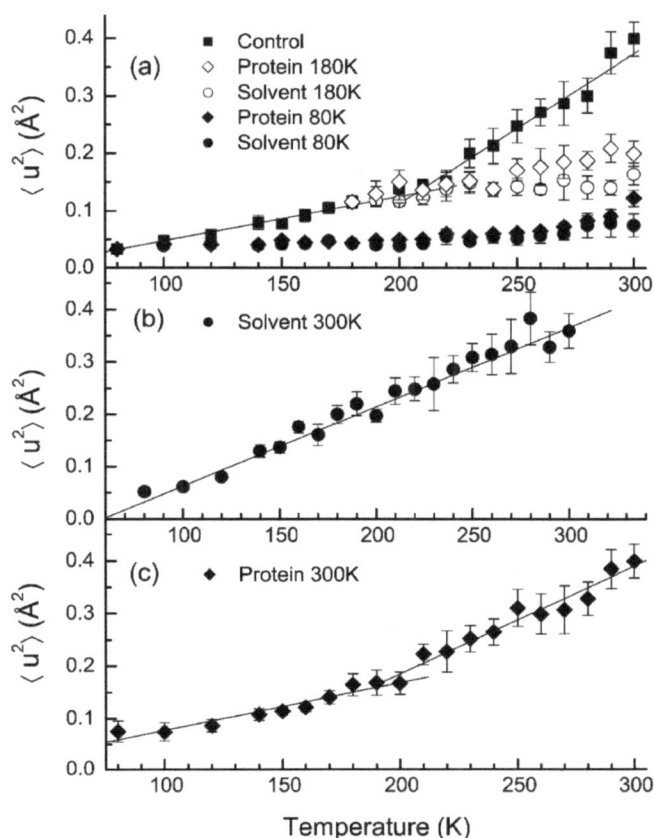

Figure 3: Mean-square fluctuations of the protein nonhydrogen atoms for different sets of simulations. (a) ■, control set with protein and solvent at same temperature; ♦, protein held at 80 K; ●, solvent held at 80 K; ◊, protein held at 180 K; ○, solvent held at 180 K. (b) Solvent held 300 K. (c) Protein held 300 K. Figure reproduced from [18].

However, there is again no clear deviation from linearity, *i.e.*, no dynamical transition behavior. In contrast, fixing the protein at 300 K and varying the solvent temperature (Fig. **1c**) recovers dynamical transition behavior in the protein, incipient at ~200 K, a slightly lower temperature than in the control set.

When fixing the solvent at 300 K, only effects due to changes with temperature in the sampled region of the protein energy landscape appear. The absence of a dynamical transition indicates, then, that these changes do not control the transition. However, when the protein is held at 300 K, variations with temperature in the sampled solvent landscape trigger the protein transition.

Fig. **4** shows the side-chain fluctuations in the control simulations as a function of distance from the protein center of mass. The dynamical transition is seen to be most pronounced in the outer parts of the protein, *i.e.*, those close to the solvent shell—above the transition the outer shells exhibit both stronger fluctuations and a larger change in gradient (inset, Fig. **4**) than the inner atoms. The solvent transition drives dynamical transition behavior primarily in the side-chain atoms of the external protein regions, *i.e.*, those closest to the solvent.

Experimental work on protonic conductivity of protein powders is consistent with a two-dimensional percolation transition of hydration water at the surfaces of various proteins upon increasing of the hydration level [27]. Computer simulations have shown that, upon increasing the hydration level, water molecules form a spanning hydrogen-bonded network enveloping protein [38-42]. Formation of this water network may play a role in collective dynamics, as a hydrogen-bonded network of water molecules may in principle exhibit dynamics which is not present in disconnected groups.

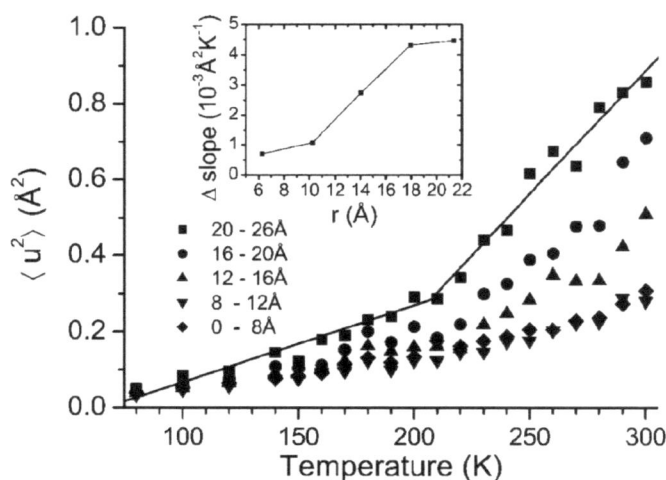

Figure 4: Mean-square fluctuations of the protein side-chain heavy atoms for five different shells, each 4 Å thick (except for the inner shell (8 Å) and outer shell (6 Å)). The inset shows the difference in slopes of lines fitted below and above 220 K as a function of distance from the protein center of mass. Linear fits to the data above and below 220 K are also shown for the outermost shell. Figure reproduced from [18].

Thermothilic Proteins

The temperature dependence of the dynamics of mesophilic and thermophilic dihydrofolate reductase has been examined using elastic incoherent neutron scattering [43]. It was demonstrated that the distribution of atomic displacement amplitudes can be derived from the elastic scattering data by assuming a (Weibull) functional form that resembles distributions seen in molecular dynamics simulations. The thermophilic enzyme has found to have a significantly broader distribution than its mesophilic counterpart. Furthermore, although the rate of increase with temperature of the atomic mean-square displacements extracted from the dynamic structure factor was found to be comparable for both enzymes, the amplitudes were found to be slightly larger for the thermophilic enzyme. Therefore, these results imply that the thermophilic enzyme is the more flexible of the two.

Protein – Protein Interactions

A physical characterization of protein-protein interactions is very important for understanding the mechanics of cell function. Recently reported all-atom lattice-dynamical calculations for a crystalline protein, ribonuclease A, showed that the sound velocities, density of states, heat capacity (C_V) and thermal diffuse scattering are all consistent with available experimental data [44]. C_V was found to be proportional to $T^{-1.68}$ for T < 35 K, significantly deviating from a Debye solid. In the vicinity of Bragg peak, inelastic scattering of X-rays by phonons was found to originate from acoustic mode scattering.

Furthermore, interprotein motions in low and fully hydrated carboxymyoglobin crystals were investigated using molecular dynamics simulation [45]. Below ≈ 240 K, the calculated dynamic structure factor showed a peak arising from interprotein vibration. Above ≈ 240 K, the intermolecular fluctuations of the fully hydrated crystal increase drastically, whereas the low-hydration model exhibits no transition. Autocorrelation function analysis demonstrated the transition to be dominated by the activation of diffusive intermolecular motion. The potential of mean force for the interaction remains quasiharmonic. Above 240 K, the intermolecular fluctuations of the fully hydrated protein crystal increase drastically, whereas a low-hydration model exhibits no transition [45]. Taken together, results above suggest an approach to protein crystal physics combining all-atom lattice-dynamical calculations with experiments on next-generation neutron sources.

Density of Protein Hydration Shell

We now present an example of the use of computational simulation to interpret experimental SANS data. Several studies have indicated that it is necessary to take into account hydration effects in SANS studies [46-48]. MD simulation provides an interpretation of neutron solution scattering data in terms of the density of water on the surface of lysozyme [46]. The simulation-derived scattering profiles are in excellent agreement with the experiment (Fig. **5**). In the simulation, the 3-Å-thick first hydration layer was found to be 15% denser than bulk water. About two-thirds of this increase is the result of a geometric contribution that would also be present if the water was unperturbed from the bulk. The remaining third arises from modification of the water structure and dynamics. It was found that hydration water density is correlated with the local electric field on the surface generated by the protein atoms as well as the local topography of the protein.

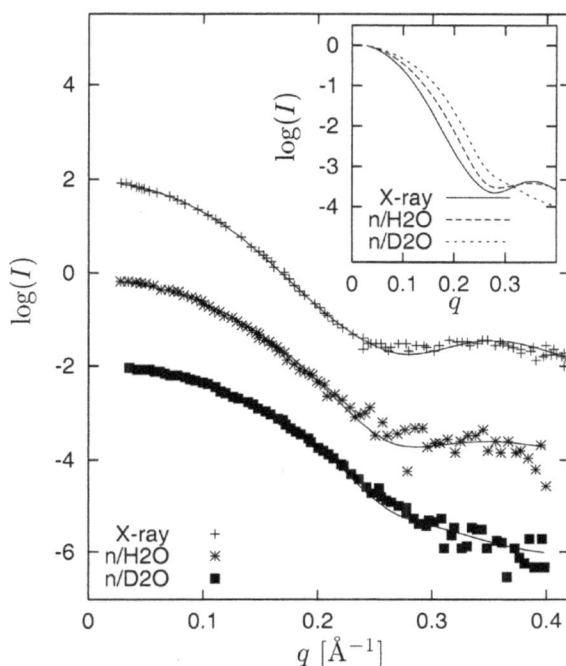

Figure 5: Comparison of MD calculated (solid lines) with experimental x-ray and neutron SAS profiles (taken from ref. [46]). In the main figure the y axis of different profiles is shifted for clarity. Inset shows calculated SAS intensities with common origin.

Structural Analysis of Lignocellulose

SANS is especially well suited for the study of the configurational statistics of disordered polymers. The experiments can be interpreted by treating the disordered polymers as fractals [49]. The analogy with fractals originates from the experimental observation that the scattering intensity I(Q) often has a power-law dependence on the scattering vector, Q:

$$I(Q) \propto Q^{-\alpha}, \tag{4}$$

where, as we will see, useful information about the system can be deduced from the value of the power-law exponent α.

When discussing SANS, it is important to consider two types of fractals: mass fractals and surface fractals [50]. The exponent, α, in Eq. (4) is directly related to the fractal dimension of the biomolecule. If $1 < \alpha < 3$, then mass-fractal scattering is observed and α is equal to the mass-fractal dimension, D_m. If $3 < \alpha < 4$, then surface scattering is observed and now $\alpha = 6 - D_s$, where D_s is the surface fractal dimension.

The above two fractal dimensions provide examples of derived quantities that can be used to compare simulation with experiments. D_s and D_m can be obtained experimentally *via* Eq. (4). As mentioned in the previous paragraph it is straight-forward to relate α to D_s and D_m.

Another way to obtain values for the fractal dimensions of biopolymers is through analysis of the configurational information from MD trajectories, *i.e.*, without reference to SANS spectra. Polymer theory links the radius of gyration of a polymer with the total number of monomers N comprising the polymer [49]:

$$R_g \propto N^{1/D_m} .$$

(5)

providing a way to determine the mass-fractal dimension D_m. To obtain the surface-fractal dimension one computes the Surface Accessible Surface Area (SASA). This is done by rolling a sphere of radius r on the surface of the molecule and then using the points visited by the center of the sphere to define a surface. The observed surface area is a function of the size of the probe radius, as using a smaller probe detects more surface details leading to a larger surface area. For a surface fractal covered with N such spheres of radius r the SASA is given by [51]:

$$A(r) \propto N \cdot r^2 \propto r^{-D_s+2} ,$$

(6)

providing a way to determine the surface-fractal dimension, D_s.

At present research is being conducted using the mass fractal and surface fractal approaches to understand lignocellulosic biomass. The recalcitrance of lignocellulosic biomass to hydrolysis is the bottleneck in the production of second-generation biofuel. Lignocellulose is a complex biomaterial made of cellulose microfibrils embedded in a matrix of polysaccharides (hemicellulose and pectins) and lignin [52], the structural analysis of which requires characterization techniques capable of spanning many length scales (from angstroms to micrometers) while differentiating between the components, such as lignin, hemicellulose and cellulose. Computer simulation can potentially act as a "virtual contrast variation" technique and separate scattering contributions from the different lignocellulose components. This is acccomplished by selecting a component (cellulose, lignin or hemicellulose) of the model and calculating the SANS profiles for this component alone.

CONCLUSIONS

In this chapter, we have provided an overview of synergistic applications of various neutron scattering techniques and computer simulation to unravel atomistic details of many dynamical phenomena of physical and biological interest that occur on the subnanosecond time scale. The advent of next-generation neutron sources (such as the Spallation Neutron Source at Oak Ridge National Laboratory) together with advanced deuteration facilities and continuing rapid increase in computing power will open up new vistas for further high-resolution insights into large length-scale and long time-scale biomolecular structure and dynamics.

ACKNOWLEDGEMENTS

The research derived here was funded in part by the Genomics:GTL Program, Office of Biological and Environmental Research, U. S. Department of Energy, under the BioEnergy Science Center. The BioEnergy Science Center is a U.S. Department of Energy Bioenergy Research Center supported by the Office of Biological and Environmental Research in the DOE Office of Science. The research is also funded in part by FWP ERKP704 "Dynamic Visualization of Lignocellulose Degradation by Integration of Neutron Scattering Imaging and Computer Simulation" funded by the DOE office of Bioscience and Environmental Research. Finally, JCS acknowledges funds from the DOE ORNL Laboratory Directed Research and Development funds (grant no: Neutron Sciences 002-0507).

REFERENCES

[1]   Brooks CL, Karplus M, Pettitt BM. Proteins : A theoretical perspective of dynamics, structure and thermodynamics. New York, Wiley, 1988.
[2]   Bee M. Quasielastic neutron scattering: principles and applications in solid state chemistry, biology and materials science. Philadelphia, Adam Hilger, 1988.
[3]   Doster W, Cusack S, Petry W. Dynamic Instability of liquid-like motions in a globular protein observed by Inelastic Neutron-Scattering. Phys Rev Lett 1990; 65: 1080-3.

[4]   Doster W, Cusack S, Petry W. Dynamical transition of myoglobin revealed by Inelastic Neutron-Scattering. Nature 1989; 337: 754-6.

[5]   Smith J, Kuczera K, Karplus M. Dynamics of myoglobin - Comparison of simulation results with neutron-scattering spectra. Proc Nat Acad Sci USA 1990; 87: 1601-5.

[6]   Knapp EW, Fischer SF, Parak F. The influence of protein dynamics on Mossbauer-spectra. J Chem Phys 1983; 78: 4701-11.

[7]   Knapp EW, Fischer SF, Parak F. Protein dynamics from Mossbauer-spectra - the temperature-dependence. J Phys Chem 1982; 86: 5042-7.

[8]   Diehl M, Doster W, Petry W, Schober H. Water-coupled low-frequency modes of myoglobin and lysozyme observed by inelastic neutron scattering. Biophys J 1997; 73: 2726-32.

[9]   Reat V, Patzelt H, Ferrand M, Pfister C, Oesterhelt D, Zaccai G. Dynamics of different functional parts of bacteriorhodopsin: H-H-2 labeling and neutron scattering. Proc Nat Acad Sci USA 1998; 95: 4970-5.

[10]   Doster W, Settles M. Protein-water displacement distributions. Biochim Biophys Acta-Prot Proteom 2005; 1749: 173-86.

[11]   Reat V, Dunn R, Ferrand M, Finney JL, Daniel PM, Smith JC. Solvent dependence of dynamic transitions in protein solutions. Proc Nat Acad Sci USA 2000; 97: 9961-6.

[12]   Daniel RM, Dunn RV, Finney JL, Smith JC. The role of dynamics in enzyme activity. Ann Rev Biophys Biomol Struct 2003; 32: 69-92.

[13]   Daniel RM, Finney JL, Reat V, Dunn R, Ferrand M, Smith JC. Enzyme dynamics and activity: Time-scale dependence of dynamical transitions in glutamate dehydrogenase solution. Biophys J 1999; 77: 2184-90.

[14]   Daniel RM, Smith JC, Ferrand M, Hery S, Dunn R, Finney JL. Enzyme activity below the dynamical transition at 220 K. Biophys J 1998; 75: 2504-7.

[15]   Kurkal V, Daniel RM, Finney JL, Tehei M, Dunn RV, Smith JC. Enzyme activity and flexibility at very low hydration. Biophys J 2005; 89: 1282-7.

[16]   Brown KG, Small EW, Peticola Wl, Erfurth SC. Conformationally dependent low-frequency motions of proteins by Laser Raman Spectroscopy. Proc Nat Acad Sci USA 1972: 69: 1467-72.

[17]   Tournier AL, Smith JC. Principal components of the protein dynamical transition. Phys Rev Lett 2003; 91: 20-8.

[18]   Tournier AL, Xu JC, Smith JC. Translational hydration water dynamics drives the protein glass transition. Biophys J 2003; 85: 1871-5.

[19]   Hayward, JA, Finney JL, Daniel RM, Smith JC. Molecular dynamics decomposition of temperature-dependent elastic neutron scattering by a protein solution. Biophys J 2003; 85: 679-85.

[20]   Hayward JA, Smith JC. Temperature dependence of protein dynamics: Computer simulation analysis of neutron scattering properties. Biophys J 2002; 82: 1216-25.

[21]   Roh JH, Curtis JE, Azzam S, Novikov VN, Peral I, Chowdhuri Z, Gregory RB, Sokolov AP. Influence of hydration on the dynamics of lysozyme. Biophys J 2006; 91: 2573-88.

[22]   Roh JH, Novikov VN, Gregory RB, Curtis JE, Chowdhuri Z, Sokolov AP. Onsets of anharmonicity in protein dynamics. Phys Rev Lett 2005; 95: 3-9.

[23]   Kneller GR, Smith JC. Liquid-Like Side-Chain Dynamics in Myoglobin. J Mol Biol 1994; 242: 181-5.

[24]   Cordone L, Ferrand M, Vitrano E, Zaccai G. Harmonic behavior of trehalose-coated carbon-monoxy-myoglobin at high temperature. Biophys J 1999; 76: 1043-7.

[25]   Andrew ER, Bryant DJ, Cashell EM. Proton Magnetic-Relaxation of Proteins in the Solid-State - Molecular-Dynamics of Ribonuclease. Chem Phys Lett 1980; 69: 551-4.

[26]   Krishnan M, Kurkal-Siebert V, Smith JC. Methyl group dynamics and the onset of anharmonicity in myoglobin. J Phys Chem B 2008; 112: 5522-33.

[27]   Rupley JA, Careri G. Protein Hydration and Function. Adv Prot Chem 1991; 41: 37-172.

[28]   Kurkal V, Daniel RM, Finney JL, Tehei M, Dunn RV, Smith JC. Low frequency enzyme dynamics as a function of temperature and hydration: A neutron scattering study. Chem Phys 2005; 317: 267-73.

[29]   Kurkal-Siebert V, Daniel RM, Finney JL, Tehei M, Dunn RV, Smith JC. Enzyme hydration, activity and flexibility: A neutron scattering approach. Biophys J 2006; 89: 4387-93.

[30]   Neusius T, Daidone I, Sokolov IM, Smith JC. Subdiffusion in peptides originates from the fractal-like structure of configuration space. Phys Rev Lett 2008; 100: 18-26.

[31]   Neusius T, Sokolov IM, Smith JC. Subdiffusion in time-averaged, confined random walks. Phys Rev E 2009; 80: 1-5.

[32]   Fitter J, Lechner RE, Dencher NA. Picosecond molecular motions in bacteriorhodopsin from neutron scattering. Biophys J 1997; 73: 2126-37.

[33]   Ferrand M, Dianoux AJ, Petry W, Zaccai G. Thermal Motions and Function of Bacteriorhodopsin in Purple Membranes -

Effects of Temperature and Hydration Studied by Neutron-Scattering. Proc Nat Acad Sci USA 1993; 90: 9668-72.

[34] Paciaroni A, Cinelli S, Onori G. Effect of the environment on the protein dynamical transition: A neutron scattering study. Biophys J 2002; 83: 1157-64.

[35] Fitter J. The temperature dependence of internal molecular motions in hydrated and dry alpha-amylase: The role of hydration water in the dynamical transition of proteins. Biophys J 1999; 76: 1034-42.

[36] Teeter MM, Yamano A, Stec B, Mohanty U. On the nature of a glassy state of matter in a hydrated protein: Relation to protein function. Proc Nat Acad Sci USA 2001; 98: 11242-7.

[37] Tournier AL, Xu JC, Smith JC. Solvent caging of internal motions in myoglobin at low temperatures. Phys Chem Comm 2003; 6: 6-8.

[38] Brovchenko I, Krukau A, Smolin N, Oleinikova A, Geiger A, Winter R. Thermal breaking of spanning water networks in the hydration shell of proteins. J Chem Phys 2005; 123: 10-7.

[39] Oleinikova A, Brovchenko I, Smolin N, Krukau A, Geiger A, Winter R. Percolation transition of hydration water: From planar hydrophilic surfaces to proteins. Phys Rev Lett 2005; 95: 4-11.

[40] Oleinikova A, Smolin N, Brovchenko I. Origin of the dynamic transition upon pressurization of crystalline proteins. J Phys Chem B 2006; 110: 19619-24.

[41] Oleinikova A, Smolin N, Brovchenko I. Influence of water clustering on the dynamics of hydration water at the surface of a lysozyme. Biophys J 2007; 93: 2986-3000.

[42] Smolin N, Oleinikova A, Brovchenko I, Geiger A, Winter R. Properties of spanning water networks at protein surfaces. J Phys Chem B 2005; 109: 10995-11005.

[43] Meinhold L, Clement D, Tehei M, Daniel R, Finney JL, Smith JC. Protein dynamics and stability: The distribution of atomic fluctuations in thermophilic and mesophilic dihydrofolate reductase derived using elastic incoherent neutron scattering. Biophys J 2008; 94: 4812-8.

[44] Meinhold L, Merzel F, Smith JC. Lattice dynamics of a protein crystal. Phys Rev Lett 2007; 99: 13-8.

[45] Kurkal-Siebert V, Agarwal R, Smith JC. Hydration-dependent dynamical transition in protein: Protein interactions at approximate to 240 k. Phys Rev Lett 2008; 100: 13-9.

[46] Merzel F, Smith JC. Is the first hydration shell of lysozyme of higher density than bulk water? Proc Nat Acad Sci USA 2002; 99: 5378-83.

[47] Hubbard SR, Hodgson KO, Doniach S. Small-Angle X-Ray-Scattering Investigation of the Solution Structure of Troponin-C. J Biol Chem 1988; 263: 4151-8.

[48] Grossmann JG, Abraham ZHL, Adman ET, Neu M, Eady RR, Smith BE, Hasnain SS. X-Ray-Scattering Using Synchrotron-Radiation Shows Nitrite Reductase from Achromobacter-Xylosoxidans to Be a Trimer in Solution. Biochem 1993; 32: 7360-6.

[49] Schmidt PW. Small-angle scattering studies of disordered, porous and fractal systems. In Munksgaard Int Publ Ltd. 1991; 414-35.

[50] Mandelbrot BB. The fractal geometry of nature. New York, Freeman WH and Co., 1983.

[51] Pfeifer P, Avnir D. Chemistry in noninteger dimensions between 2 and 3. 1. Fractal theory of heterogeneous surfaces. J Chem Phys 1983; 79: 3558-65.

[52] Cosgrove DJ. Growth of the plant cell wall. Nat Rev Mol Cell Biol 2005; 6: 850-61.

AUTHOR INDEX

SUBJECT INDEX

M

Membranes 36,79
Mean square displacement 8,22,37,47,87,101
Myelin 37
Myoglobin 22,47,67,92,101

P

Polypeptides 92

Q

Quasi-Elastic Neutron Scattering 4,36,102

R

Raman Spectroscopy 86
Relaxation 17,24,47,67,101
Ribonuclease A 16

S

Saccharides 22,66,79
Self-Distribution-Function Procedure 22
Simulation 66,99
Small Angle Neutron Scattering 11,99
SNase 90

V

Vibrational motions 4,26,36,67,86

W

Water 5,22,38,47,66,79,86,102

www.ingramcontent.com/pod-product-compliance
Lightning Source LLC
Chambersburg PA
CBHW041719210326
41598CB00007B/703